WORKING GUIDE TO
VAPOR-LIQUID PHASE
EQUILIBRIA CALCULATIONS

WORKING GUIDE TO VAPOR-LIQUID PHASE EQUILIBRIA CALCULATIONS

TAREK AHMED

ELSEVIER

AMSTERDAM • BOSTON • HEIDELBERG • LONDON
NEW YORK • OXFORD • PARIS • SAN DIEGO
SAN FRANCISCO • SINGAPORE • SYDNEY • TOKYO
Gulf Professional Publishing is an imprint of Elsevier

Gulf Professional Publishing is an imprint of Elsevier
30 Corporate Drive, Suite 400, Burlington, MA 01803, USA
Linacre House, Jordan Hill, Oxford OX2 8DP, UK

Library of Congress Cataloging-in-Publication Data
A catalog record for this book is available from the Library of Congress.

British Library Cataloguing-in-Publication Data
A catalogue record for this book is available from the British Library.

ISBN: 978-1-85617-826-6

For information on all Gulf Professional Publishing publications
visit our Web site at www.elsevierdirect.com

09 10 11 12 13 10 9 8 7 6 5 4 3 2 1

Printed in the United States of America

Working together to grow libraries in developing countries

www.elsevier.com | www.bookaid.org | www.sabre.org

ELSEVIER   BOOK AID International   Sabre Foundation

Contents

6

APPLICATIONS OF THE EQUILIBRIUM RATIO IN RESERVOIR ENGINEERING

7

EQUATIONS OF STATE

8

APPLICATIONS OF THE EQUATION OF STATE IN PETROLEUM ENGINEERING

9

SPLITTING AND LUMPING SCHEMES OF THE PLUS FRACTION

Vapor Pressure

A phase is defined as the part of a system that is uniform in physical and chemical properties, homogeneous in composition, and separated from other coexisting phases by definite boundary surfaces. The most important phases occurring in petroleum production are the hydrocarbon liquid phase and the gas phase. Water is also commonly present as an additional liquid phase. These can coexist in equilibrium when the variables describing change in the entire system remain constant over time and position. The chief variables that determine the state of equilibrium are system temperature, system pressure, and composition.

The conditions under which these different phases can exist are a matter of considerable practical importance in designing surface separation facilities and developing compositional models. These types of calculations are based on the concept of equilibrium ratios.

A system that contains only one component is considered the simplest type of hydrocarbon system. The word *component* refers to the number of molecular or atomic species present in the substance. A single-component system is composed entirely of one kind of atom or molecule. We often use the word *pure* to describe a single-component system. The qualitative understanding of the relationship that exists between temperature T, pressure p, and volume V of pure components can provide an excellent basis for understanding the phase behavior of complex hydrocarbon mixtures.

Consider a closed evacuated container that has been partially filled with a pure component in the liquid state. The molecules of the liquid are in constant motion with different velocities. When one of these molecules reaches the liquid surface, it may possess sufficient kinetic energy to overcome the attractive forces in the liquid and pass into the vapor spaces above. As the number of molecules in the vapor phase increases, the rate of return to the liquid phase also increases. A state of equilibrium is eventually reached when the number of molecules leaving and returning is equal. The molecules in the vapor phase obviously exert a pressure on the wall of the container, and this pressure is defined as the vapor pressure, p_v. As the temperature of the liquid increases, the average molecular velocity increases, with a larger number of molecules possessing sufficient energy to enter the vapor phase. As a result, the vapor pressure of a pure component in the liquid state increases with increasing temperature.

A method that is particularly convenient for expressing the vapor pressure of pure substances as a function of temperature is shown in Figure 1-1. The chart, known as the *Cox chart*, uses a logarithmic scale

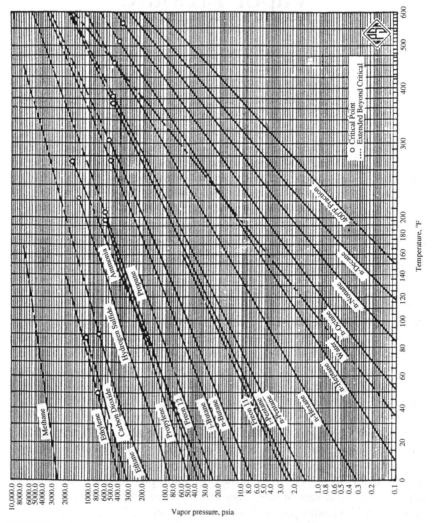

FIGURE 1-1 Vapor pressures for hydrocarbon components. (Courtesy of the Gas Processors Suppliers Association, *Engineering Data Book*, 10th Ed., 1987.)

for the vapor pressure and an entirely arbitrary scale for the temperature in °F. The vapor pressure curve for any particular component, as shown in Figure 1-1, can be defined as the dividing line between the area where vapor and liquid exist. If the system pressure exists at its vapor pressure, two phases can coexist in equilibrium. Systems represented by points located below that vapor pressure curve are composed of only the vapor phase. Similarly, points above the curve represent systems that exist in the liquid phase. These statements can be conveniently summarized by the following expressions:

- $p < p_v \rightarrow$ system is entirely in the vapor phase.
- $p > p_v \rightarrow$ system is entirely in the liquid phase.
- $p = p_v \rightarrow$ vapor and liquid coexist in equilibrium.

where p is the pressure exerted on the pure component. Note that these expressions are valid only if the system temperature T is below the critical temperature T_c of the substance.

The vapor pressure chart allows a quick determination of the p_v of a pure component at a specific temperature. For computer and spreadsheet applications, however, an equation is more convenient. Lee and Kesler (1975) [29] proposed the following generalized vapor pressure equation:

$$p_v = p_c \exp(A + \omega B)$$

with

$$A = 5.92714 - \frac{6.09648}{T_r} - 1.2886 \ln(T_r) + 0.16934(T_r)^6$$

$$B = 15.2518 - \frac{15.6875}{T_r} - 13.4721 \ln(T_r) + 0.4357(T_r)^6$$

where p_v = vapor pressure, psi
p_c = critical pressure, psi
T_r = reduced temperature (T / T_c)
T = system temperature, °R
T_c = critical temperature, °R
ω = acentric factor

Equilibrium Ratios

In a multicomponent system, the equilibrium ratio K_i of a given component is defined as the ratio of the mole fraction of the component in the gas phase y_i to the mole fraction of the component in the liquid phase x_i. Mathematically, the relationship is expressed as

$$K_i = \frac{y_i}{x_i} \tag{2-1}$$

where K_i = equilibrium ratio of component i
$\quad y_i$ = mole fraction of component i in the gas phase
$\quad x_i$ = mole fraction of component i in the liquid phase

At pressures below 100 psia, Raoult's and Dalton's laws for ideal solutions provide a simplified means of predicting equilibrium ratios. Raoult's law states that the partial pressure p_i of a component in a multicomponent system is the product of its mole fraction in the liquid phase x_i and the vapor pressure of the component p_{vi}, or

$$p_i = x_i p_{vi} \tag{2-2}$$

where p_i = partial pressure of component i, psia
$\quad p_{vi}$ = vapor pressure of component i, psia
$\quad x_i$ = mole fraction of component i in the liquid phase

Dalton's law states that the partial pressure of a component is the product of its mole fraction in the gas phase y_i and the total pressure of the system p, or

$$p_i = y_i p \tag{2-3}$$

where p = total system pressure, psia.

At equilibrium and in accordance with the preceding laws, the partial pressure exerted by a component in the gas phase must be equal to the

5

partial pressure exerted by the same component in the liquid phase. Therefore, equating the equations describing the two laws yields

$$x_i p_{vi} = y_i p$$

Rearranging this relationship and introducing the concept of the equilibrium ratio gives

$$\frac{y_i}{x_i} = \frac{p_{vi}}{p} = K_i \qquad (2\text{-}4)$$

Equation 2-4 shows that, for ideal solutions and regardless of the overall composition of the hydrocarbon mixture, the equilibrium ratio is only a function of the system pressure p and the temperature T since the vapor pressure of a component is only a function of temperature (see Figure 1-1).

It is appropriate at this stage to introduce and define the following nomenclatures:

z_i = mole fraction of component in the entire hydrocarbon mixture
n = total number of moles of the hydrocarbon mixture, lb-mol
n_L = total number of moles in the liquid phase
n_v = total number of moles in the vapor (gas) phase

By definition,

$$n = n_L + n_v \qquad (2\text{-}5)$$

Equation 2-5 indicates that the total number of moles in the system is equal to the total number of moles in the liquid phase plus the total number of moles in the vapor phase. A material balance on the ith component results in

$$z_i n = x_i n_L + y_i n_v \qquad (2\text{-}6)$$

where $z_i n$ = total number of moles of component i in the system
$x_i n_L$ = total number of moles of component i in the liquid phase
$y_i n_v$ = total number of moles of component i in the vapor phase

Also by the definition of mole fraction, we may write

$$\sum_i z_i = 1 \qquad (2\text{-}7)$$

$$\sum_i x_i = 1 \qquad (2\text{-}8)$$

$$\sum_i y_i = 1 \qquad (2\text{-}9)$$

It is convenient to perform all phase-equilibria calculations on the basis of 1 mol of the hydrocarbon mixture, i.e., n = 1. That assumption reduces Equations 2-5 and 2-6 to

$$n_L + n_v = 1 \tag{2-10}$$

$$x_i n_L + y_i n_v = z_i \tag{2-11}$$

Combining Equations 2-4 and 2-11 to eliminate y_i from Equation 2-11 gives

$$x_i\, n_L + (x_i K_i) n_v = z_i$$

Solving for x_i yields

$$x_i = \frac{z_i}{n_L + n_v\, K_i} \tag{2-12}$$

Equation 2-11 can also be solved for y_i by combining it with Equation 2-4 to eliminate x_i:

$$y_i = \frac{z_i K_i}{n_L + n_v K_i} = x_i K_i \tag{2-13}$$

Combining Equation 2-12 with 2-8 and Equation 2-13 with 2-19 results in

$$\sum_i x_i = \sum_i \frac{z_i}{n_L + n_v K_i} = 1 \tag{2-14}$$

$$\sum_i y_i = \sum_i \frac{z_i K_i}{n_L + n_v K_i} = 1 \tag{2-15}$$

Since

$$\sum_i y_i - \sum_i x_i = 0$$

therefore,

$$\sum_i \frac{z_i K_i}{n_L + n_v K_i} - \sum_i \frac{z_i}{n_L + n_v K_i} = 0$$

or

$$\sum_i \frac{z_i(K_i - 1)}{n_L + n_v K_i} = 0$$

Replacing n_L with $(1 - n_v)$ yields

$$f(n_v) = \sum_i \frac{z_i(K_i - 1)}{n_v(K_i - 1) + 1} = 0 \tag{2-16}$$

This set of equations provides the necessary phase relationships to perform volumetric and compositional calculations on a hydrocarbon system. These calculations are referred to as *flash calculations* and are discussed in Chapter 3.

3

Flash Calculations

Flash calculations are an integral part of all reservoir and process engineering calculations. They are required whenever it is desirable to know the amounts (in moles) of hydrocarbon liquid and gas coexisting in a reservoir or a vessel at a given pressure and temperature. These calculations are also performed to determine the composition of the existing hydrocarbon phases.

Given the overall composition of a hydrocarbon system at a specified pressure and temperature, flash calculations are performed to determine

- Moles of the gas phase n_v.
- Moles of the liquid phase n_L.
- Composition of the liquid phase x_i.
- Composition of the gas phase y_i.

The computational steps for determining n_L, n_v, y_i, and x_i of a hydrocarbon mixture with a known overall composition of z_i and characterized by a set of equilibrium ratios K_i are summarized as follows:

Step 1. **Calculation of n_v.** Equation 2-16 can be solved for n_v using the Newton-Raphson iteration technique. In applying this iterative technique,

- Assume any arbitrary value of n_v between 0 and 1, e.g., $n_v = 0.5$. A good assumed value may be calculated from the following relationship, providing that the values of the equilibrium ratios are accurate:

$$n_v = A/(A - B)$$

where

$$A = \sum_i [z_i(K_i - 1)]$$
$$B = \sum_i [z_i(K_i - 1)/K_i]$$

- Evaluate the function $f(n_v)$ as given by Equation 2-16 using the assumed value of n_v.
- If the absolute value of the function $f(n_v)$ is smaller than a preset tolerance, e.g., 10^{-15}, then the assumed value of n_v is the desired solution.
- If the absolute value of $f(n_v)$ is greater than the preset tolerance, then a new value of n_v is calculated from the following expression:

$$(n_v)_n = n_v - f(n_v)/f'(n_v)$$

with

$$f' = -\sum_i \left\{ \frac{z_i(K_i - 1)^2}{[n_v(K_i - 1) + 1]^2} \right\}$$

where $(n_v)_n$ is the new value of n_v to be used for the next iteration.
- This procedure is repeated with the new values of n_v until convergence is achieved.

Step 2. **Calculation of n_L.** Calculate the number of moles of the liquid phase from Equation 2-10, to give

$$n_L = 1 - n_v$$

Step 3. **Calculation of x_i.** Calculate the composition of the liquid phase by applying Equation 2-12:

$$x_i = \frac{z_i}{n_L + n_v K_i}$$

Step 4. **Calculation of y_i.** Determine the composition of the gas phase from Equation 3-13:

$$y_i = \frac{z_i K_i}{n_L + n_v K_i} = x_i K_i$$

Example 3-1

A hydrocarbon mixture with the following overall composition is flashed in a separator at 50 psia and 100°F.

Component	z_i
C_3	0.20
$i - C_4$	0.10
$n - C_4$	0.10

i – C_5	0.20
n – C_5	0.20
C_6	0.20

Assuming an ideal solution behavior, perform flash calculations.

Solution

Step 1. Determine the vapor pressure from the Cox chart (Figure 1-1) and calculate the equilibrium ratios from Equation 2-4:

Component	z_i	p_{vi} at 100°F	$K_i = p_{vi}/50$
C_3	0.20	190	3.80
i – C_4	0.10	72.2	1.444
n – C_4	0.10	51.6	1.032
i – C_5	0.20	20.44	0.4088
n – C_5	0.20	15.57	0.3114
C_6	0.20	4.956	0.09912

Step 2. Solve Equation 2-16 for n_v using the Newton-Raphson method, to give

Iteration	n_v	$f(n_v)$
0	0.08196579	3.073 E-02
1	0.1079687	8.894 E-04
2	0.1086363	7.60 E-07
3	0.1086368	1.49 E-08
4	0.1086368	0.0

Step 3. Solve for n_L:

$$n_L = 1 - n_v$$
$$n_L = 1 - 0.1086368 = 0.8913631$$

Step 4. Solve for x_i and y_i to yield

Component	z_i	K_i	$x_i = z_i/(0.8914+0.1086\,K_i)$	$y_i = x_i K_i$
C_3	0.20	3.80	0.1534	0.5829
i – C_4	0.10	1.444	0.0954	0.1378
n – C_4	0.10	1.032	0.0997	0.1029
i – C_5	0.20	0.4088	0.2137	0.0874
n – C_5	0.20	0.3114	0.2162	0.0673
C_6	0.20	0.09912	0.2216	0.0220

Notice that, for a binary system, i.e., a two-component system, flash calculations can be performed without resorting to the iterative technique by applying the following steps:

Step 1. **Solve for the composition of the liquid phase x_i.** From Equations 2-8 and 2-9:

$$\sum_i x_i = x_1 + x_2 = 1$$
$$\sum_i y_i = y_1 + y_2 = K_1 x_1 + K_2 x_2 = 1$$

Solving these two expressions for the liquid compositions x_1 and x_2 gives

$$x_1 = \frac{1 - K_2}{K_1 - K_2}$$

and

$$x_2 = 1 - x_1$$

where x_1 = mole fraction of the first component in the liquid phase
x_2 = mole fraction of the second component in the liquid phase
K_1 = equilibrium ratio of the first component
K_2 = equilibrium ratio of the second component

Step 2. **Solve for the composition of the gas phase y_i.** From the definition of the equilibrium ratio, calculate the composition of the gas as follows:

$$y_1 = x_1 K_1$$
$$y_2 = x_2 K_2 = 1 - y_1$$

Step 3. **Solve for the number of moles of the vapor phase n_v.** Arrange Equation 2-12 to solve for n_v, to give

$$n_v = \frac{z_1 - x_1}{x_1(K_1 - 1)}$$

and

$$n_l = 1 - n_v$$

where
z_1 = mole fraction of the first component in the entire system
x_1 = mole fraction of the first component in the liquid phase
K_1 = equilibrium ratio of the first component
K_2 = equilibrium ratio of the second component

Equilibrium Ratios for Real Solutions

The equilibrium ratios, which indicate the partitioning of each component between the liquid phase and the gas phase, as calculated by Equation 2-4 in terms of vapor pressure and system pressure, proved to be inadequate. The basic assumptions behind Equation 2-4 are that

- The vapor phase is an ideal gas as described by Dalton's law.
- The liquid phase is an ideal solution as described by Raoult's law.

This combination of assumptions is unrealistic and results in inaccurate predictions of equilibrium ratios at high pressures.

For a real solution, the equilibrium ratios are no longer a function of the pressure and temperature alone but are also a function of the composition of the hydrocarbon mixture. This observation can be stated mathematically as

$$K_i = K(p, T, z_i)$$

Numerous methods have been proposed for predicting the equilibrium ratios of hydrocarbon mixtures. These correlations range from a simple mathematical expression to a complicated expression containing several composition-dependent variables. The following methods are presented:

- Wilson's correlation.
- Standing's correlation.
- Convergence pressure method.
- Whitson and Torp correlation.

4.1 WILSON'S CORRELATION

Wilson (1968) [67] proposed a simplified thermodynamic expression for estimating K values. The proposed expression has the following form:

$$K_i = \frac{p_{ci}}{p} \exp\left[5.37(1 + \omega_i)\left(1 - \frac{T_{ci}}{T}\right)\right] \tag{4-1}$$

where p_{ci} = critical pressure of component i, psia
     p = system pressure, psia
     T_{ci} = critical temperature of component i, °R
     T = system temperature, °R
     ω_i = acentric factor of component i

This relationship generates reasonable values for the equilibrium ratio when applied at low pressures.

4.2 STANDING'S CORRELATION

Hoffmann et al. (1953) [6], Brinkman and Sicking (1960) [11], Kehn (1964) [21], and Dykstra and Mueller (1965) [27] suggested that any pure hydrocarbon or nonhydrocarbon component could be uniquely characterized by combining its boiling-point temperature, critical temperature, and critical pressure into a characterization parameter that is defined by the following expression:

$$F_i = b_i[1/T_{bi} - 1/T] \tag{4-2}$$

with

$$b_i = \frac{\log(p_{ci}/14.7)}{[1/T_{bi} - 1/T_{ci}]} \tag{4-3}$$

where F_i = component characterization factor
     T_{bi} = normal boiling point of component i, °R

Standing (1979) [57] derived a set of equations that fit the equilibrium ratio data of [24] at pressures of less than 1000 psia and temperatures below 200 °F. The proposed form of the correlation is based on an observation that plots of $\log(K_i p)$ versus F_i at a given pressure often form straight lines. The basic equation of the straight-line relationship is given by

$$\log(K_i p) = a + cF_i$$

Solving for the equilibrium ratio K_i gives

$$K_i = \frac{1}{p}10^{(a+cF_i)} \tag{4-4}$$

where the coefficients a and c are the intercept and the slope of the line, respectively.

From a total of six isobar plots of $\log(K_i p)$ versus F_i for 18 sets of equilibrium ratio values, Standing correlated the coefficients a and c with the pressure, to give

$$a = 1.2 + 0.00045p + 15(10^{-8})p^2 \tag{4-5}$$

$$c = 0.89 - 0.00017p - 3.5(10^{-8})p^2 \tag{4-6}$$

Standing pointed out that the predicted values of the equilibrium ratios of N_2, CO_2, H_2S, and C_1 through C_6 can be improved considerably by changing the correlating parameter b_i and the boiling point of these components. The author proposed the following modified values:

Component	b_i	T_{bi} °R
N_2	470	109
CO_2	652	194
H_2S	1136	331
C_1	300	94
C_2	1145	303
C_3	1799	416
$i - C_4$	2037	471
$n - C_4$	2153	491
$i - C_5$	2368	542
$n - C_5$	2480	557
C_6*	2738	610
$n - C_6$	2780	616
$n - C_7$	3068	669
$n - C_8$	3335	718
$n - C_9$	3590	763
$n - C_{10}$	3828	805

*Lumped hexanes-fraction.

When making flash calculations, the question of the equilibrium ratio to use for the lumped heptanes-plus fraction always arises. One rule of thumb proposed by Katz and Hachmuth (1937) [24] is that the K value for C_{7+} can be taken as 15% of the K of C_7, or

$$K_{C_{7+}} = 0.15 \, K_{C_{7+}}$$

Standing offered an alternative approach for determining the K value of the heptanes and heavier fractions. By imposing experimental equilibrium ratio values for C_{7+} on Equation 4-4, Standing calculated the corresponding characterization factors F_i for the plus fraction. The calculated F_i values were used to specify the pure normal paraffin hydrocarbon having the K value of the C_{7+} fraction.

Standing suggested the following computational steps for determining the parameters b and T_b of the heptanes-plus fraction.

Step 1. Determine, from the following relationship, the number of carbon atoms n of the normal paraffin hydrocarbon having the K value of the C_{7+} fraction:

$$n = 7.30 + 0.0075(T - 460) + 0.0016p \qquad (4\text{-}7)$$

Step 2. Calculate the correlating parameter b and the boiling point T_b from the following expressions:

$$b = 1{,}013 + 324n - 4.256n^2 \qquad (4\text{-}8)$$

$$T_b = 301 + 59.85n - 0.971n^2 \qquad (4\text{-}9)$$

These calculated values can then be used in Equation 4-2 to evaluate F_i for the heptanes-plus fraction, i.e., $F_{C_{7+}}$. It is also interesting to note that experimental phase equilibria data suggest that the equilibrium ratio for carbon dioxide can be closely approximated by the following relationship:

$$K_{CO_2} = \sqrt{K_{C_1} K_{C_2}}$$

where K_{CO_2} = equilibrium ratio of CO_2
      K_{C_1} = equilibrium ratio of methane
      K_{C_2} = equilibrium ratio of ethane

Example 4-1

A hydrocarbon mixture with the following composition is flashed at 1000 psia and 150 °F:

Component	z_i
CO_2	0.009
N_2	0.003
C_1	0.535
C_2	0.115
C_3	0.088
$i - C_4$	0.023
$n - C_4$	0.023
$i - C_5$	0.015
$n - C_5$	0.015
C_6	0.015
C_{7+}	0.159

If the molecular weight and specific gravity of C_{7+} are 150.0 and 0.78, respectively, calculate the equilibrium ratios by using

a. Wilson's correlation.
b. Standing's correlation.

Solution

a. Wilson's Correlation

Step 1. Calculate the critical pressure, critical temperature, and acentric factor of C_{7+} using the characterization method of Riazi and Daubert (1987):

$$T_c = 1139.4°R, \ p_c = 320.3 \text{ psia}, \ \omega = 0.5067$$

Step 2. Apply Equation 4-1 to give

				$K_i = \dfrac{P_{ci}}{1000}$ $\exp\left[5.37(1+\omega_1)\left(1 - \dfrac{T_{ci}}{610}\right)\right]$
Component	P_c, psia	T_c, °R	ω	
CO_2	1071	547.9	0.225	2.0923
N_2	493	227.6	0.040	16.343
C_1	667.8	343.37	0.0104	7.155
C_2	707.8	550.09	0.0986	1.236
C_3	616.3	666.01	0.1542	0.349
$i - C_4$	529.1	734.98	0.1848	0.144
$n - C_4$	550.7	765.65	0.2010	0.106
$i - C_5$	490.4	829.1	0.2223	0.046
$n - C_5$	488.6	845.7	0.2539	0.036
C_6	436.9	913.7	0.3007	0.013
C_{7+}	320.3	1139.4	0.5069	0.00029

b. Standing's Correlation

Step 1. Calculate coefficients a and c from Equations 4-5 and 4-6 to give

$$a = 1.2 + 0.00045(1000) + 15(10^{-8})(1000)^2 = 1.80$$
$$c = 0.89 - 0.00017(1000) - 3.5(10^{-8})(1000)^2 = 0.685$$

Step 2. Calculate the number of carbon atoms n from Equation 4-7 to give

$$n = 7.3 + 0.0075(150) + 0.0016(1000) = 10.025$$

Step 3. Determine the parameter b and the boiling point T_b for the hydrocarbon component with n carbon atoms using Equations 4-8 and 4-9 to yield

$$b = 1013 + 324(10.025) - 4.256(10.025)^2 = 3833.369$$
$$T_b = 301 + 59.85(10.025) - 0.971(10.025)^2 = 803.41°R$$

Step 4. Apply Equation 4-4, to give

Component	b_i	T_{bi}	F_i, Eq. 4-2	K_i, Eq. 4-4
CO_2	652	194	2.292	2.344
N_2	470	109	3.541	16.811
C_1	300	94	2.700	4.462
C_2	1145	303	1.902	1.267
C_3	1799	416	1.375	0.552
$i - C_4$	2037	471	0.985	0.298
$n - C_4$	2153	491	0.855	0.243
$i - C_5$	2368	542	0.487	0.136
$n - C_5$	2480	557	0.387	0.116
C_6	2738	610	0	0.063
C_{7+}	3833.369	803.41	-1.513	0.0058

4.3 CONVERGENCE PRESSURE METHOD

Early high-pressure phase-equilibria studies revealed that when a hydrocarbon mixture of a fixed overall composition is held at a constant temperature as the pressure increases, the equilibrium values of all components converge toward a common value of unity at a certain pressure. This pressure is termed the convergence pressure P_k of the hydrocarbon mixture. The convergence pressure is essentially used to correlate the effect of the composition on equilibrium ratios.

The concept of convergence pressure can be better appreciated by examining Figure 4-1. The figure shows a schematic diagram of a typical set of equilibrium ratios plotted versus pressure on log-log paper for a hydrocarbon mixture held at a constant temperature. The illustration shows a tendency of the equilibrium ratios to converge isothermally to a value of $K_i = 1$ for all components at a specific pressure, i.e., the convergence pressure. A different hydrocarbon mixture may exhibit a different convergence pressure.

The Natural Gas Processors Suppliers Association (NGPSA) correlated a considerable quantity of K-factor data as a function of temperature, pressure, component identity, and convergence pressure. These correlation charts were made available through the NGPSA's *Engineering Data Book* and are considered to be the most extensive set of published equilibrium ratios for hydrocarbons. They include

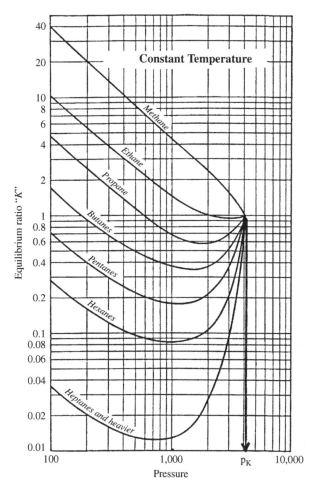

FIGURE 4-1 Equilibrium ratios for a hydrocarbon system.

the K values for a number of convergence pressures, specifically 800, 1000, 1500, 2000, 3000, 5000, and 10,000 psia. Equilibrium ratios for methane through decane and for a convergence pressure of 5000 psia are also given.

Several investigators observed that, for hydrocarbon mixtures with convergence pressures of 4000 psia or greater, the values of the equilibrium ratios are essentially the same for hydrocarbon mixtures with system pressures of less than 1000 psia. This observation led to the conclusion that the overall composition of the hydrocarbon mixture has little effect on equilibrium ratios when the system pressure is less than 1000 psia.

The problem with using the NGPSA equilibrium ratio graphical correlations is that the convergence pressure must be known before selecting
the appropriate charts. Three of the methods of determining the convergence pressure are discussed next.

Hadden's Method

Hadden (1953) [18] developed an iterative procedure for calculating the
convergence pressure of the hydrocarbon mixture. The procedure is based
on forming a "binary system" that describes the entire hydrocarbon mixture. One of the components in the binary system is selected as the lightest
fraction in the hydrocarbon system and the other is treated as a "pseudo-
component" that lumps all the remaining fractions. The binary system
concept uses the binary system convergence pressure chart, as shown in
Figure 4-2, to determine the p_k of the mixture at the specified temperature.

The equivalent binary system concept employs the following steps for
determining the convergence pressure:

Step 1. Estimate a value for the convergence pressure.

Step 2. From the appropriate equilibrium ratio charts, read the K values
of each component present in the mixture by filling in the charts
with the system pressure and temperature.

Step 3. Perform flash calculations using the calculated K values and
system composition.

Step 4. Identify the lightest hydrocarbon component that comprises
at least 0.1 mol % in the liquid phase.

Step 5. Convert the liquid mole fraction to a weight fraction.

Step 6. Exclude the lightest hydrocarbon component, as identified in step 4,
and normalize the weight fractions of the remaining components.

Step 7. Calculate the weight average critical temperature and pressure
of the lumped components (pseudo-components) from the
following expressions:

$$T_{pc} = \sum_{i=2} w_i^* T_{ci}$$

$$P_{pc} = \sum_{i=2} w_i^* P_{ci}$$

where w_i^* = normalized weight fraction of component i
T_{pc} = pseudo-critical temperature, °R
P_{pc} = pseudo-critical pressure, psi

Step 8. Enter the critical properties of the pseudo-component into
Figure 4-2 and trace the critical locus of the binary consisting
of the light component and the pseudo-component.

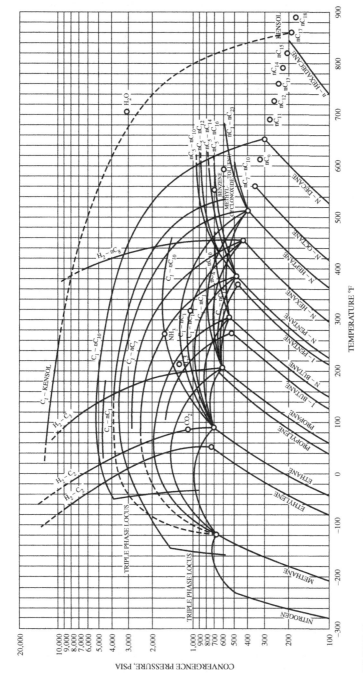

FIGURE 4.2 Convergence pressures for binary systems. (Courtesy of the Gas Processors Suppliers Association, *Engineering Data Book,* 10th Ed., 1987.)

Step 9. Read the new convergence pressure (ordinate) from the point at which the locus crosses the temperature of interest.

Step 10. If the calculated new convergence pressure is not reasonably close to the assumed value, repeat steps 2 through 9.

Note that, when the calculated new convergence pressure is between values for which charts are provided, interpolation between charts might be necessary. If the K values do not change rapidly with the convergence pressure, i.e., $p_k \gg p$, then the set of charts nearest to the calculated p_k may be used.

Standing's Method

Standing (1977) [56] suggested that the convergence pressure can be roughly correlated linearly with the molecular weight of the heptanes-plus fraction. Whitson and Torp (1981) [65] expressed this relationship by the following equation:

$$p_k = 60M_{C_{7+}} - 4200 \qquad (4\text{-}10)$$

where $M_{C_{7+}}$ is the molecular weight of the heptanes-plus fraction.

Rzasa's Method

Rzasa et al. (1952) [49] presented a simplified graphical correlation for predicting the convergence pressure of light hydrocarbon mixtures. They used the temperature and the product of the molecular weight and specific gravity of the heptanes-plus fraction as correlating parameters. The graphical illustration of the proposed correlation is shown in Figure 4-3.

The graphical correlation is expressed mathematically by the following equation:

$$p_k = -2381.8542 + 46.341487[M\gamma]_{C_{7+}} + \sum_{i=1}^{3} a_i \left[\frac{(M\gamma)_{C_{7+}}}{T - 460} \right]^i \qquad (4\text{-}11)$$

where $(M)_{C_{7+}}$ = molecular weight of C_{7+}
$(\gamma)_{C_{7+}}$ = specific gravity of C_{7+}
T = temperature, °R
$a_1 - a_3$ = coefficients of the correlation with the following values:
$a_1 = 6{,}124.3049$
$a_2 = -2{,}753.2538$
$a_3 = 415.42049$

This mathematical expression can be used for determining the convergence pressure of hydrocarbon mixtures at temperatures in the range of 50 to 300°F.

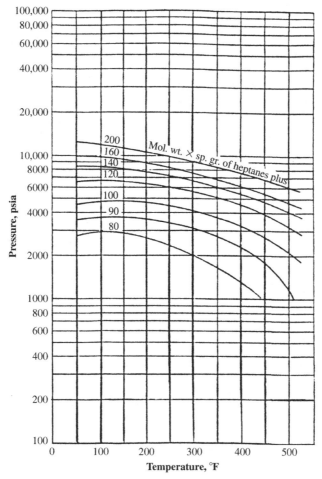

FIGURE 4-3 Rzasa's convergence pressure correlation. (Courtesy of the American Institute of Chemical Engineers.)

4.4 WHITSON AND TORP CORRELATION

Whitson and Torp (1981) [65] reformulated Wilson's equation (Equation 4-1) to yield accurate results at higher pressures. Wilson's equation was modified by incorporating the convergence pressure into the correlation, to give:

$$K_i = \left(\frac{p_{ci}}{p_k}\right)^{A-1} \left(\frac{p_{ci}}{p}\right) \exp\left[5.37A(1 + \omega_i)\left(1 - \frac{T_{ci}}{T}\right)\right] \qquad (4\text{-}12)$$

with

$$A = 1 - \left(\frac{p}{p_k}\right)^{0.7}$$

where p = system pressure, psig
    p_k = convergence pressure, psig
    T = system temperature, °R
    ω_i = acentric factor of component i

Example 4-2

Rework Example 4-1 and calculate the equilibrium ratios using the Whitson and Torp method.

Solution

Step 1. Determine the convergence pressure from Equation 4-11 to give
    $P_k = 9{,}473.89$.
Step 2. Calculate the coefficient A:

$$A = 1 - \left(\frac{1000}{9474}\right)^{0.7} = 0.793$$

Step 3. Calculate the equilibrium ratios from Equation 4-12 to give

$$K_i = \left(\frac{P_{ci}}{9474}\right)^{0.793^{-1}}$$

Component	p_c, psia	T_c, °R	ω	$\frac{P_{ci}}{1000}\exp\left[5.37A(1+\omega_i)\left(1-\frac{T_{ci}}{610}\right)\right]$
CO_2	1071	547.9	0.225	2.9
N_2	493	227.6	0.040	14.6
C_1	667.8	343.37	0.0104	7.6
C_2	707.8	550.09	0.0968	2.1
C_3	616.3	666.01	0.1524	0.7
$i-C_4$	529.1	734.98	0.1848	0.42
$n-C_4$	550.7	765.65	0.2010	0.332
$i-C_5$	490.4	829.1	0.2223	0.1749
$n-C_5$	488.6	845.7	0.2539	0.150
C_6	436.9	913.7	0.3007	0.0719
C_{7+}	320.3	1139.4	0.5069	$0.683(10^{-3})$

Equilibrium Ratios for the Plus Fraction

The equilibrium ratios of the plus fraction often behave in a manner different from the other components of a system. This is because the plus fraction in itself is a mixture of components. Several techniques have been proposed for estimating the K value of the plus fractions. Some of these techniques are presented here.

5.1 CAMPBELL'S METHOD

Campbell (1976) [7] proposed that the plot of the log of K_i versus T_{ci}^2 for each component is a linear relationship for any hydrocarbon system. Campbell suggested that, by drawing the best straight line through the points for propane through hexane components, the resulting line can be extrapolated to obtain the K value of the plus fraction. He pointed out that the plot of log K_i versus $1/T_{bi}$ of each heavy fraction in the mixture is also a straight-line relationship. The line can be extrapolated to obtain the equilibrium ratio of the plus fraction from the reciprocal of its average boiling point.

5.2 WINN'S METHOD

Winn (1954) proposed the following expression for determining the equilibrium ratio of heavy fractions with a boiling point above 210°F:

$$K_{C_+} = \frac{K_{C_7}}{(K_{C_2}/K_{C_7})^b} \tag{5-1}$$

where K_{C_+} = value of the plus fraction

K_{C_7} = K value of n-heptane at system pressure, temperature, and convergence pressure

K_{C_7} = K value of ethane

b = volatility exponent

Winn correlated, graphically, the volatility component b of the heavy fraction, with the atmosphere boiling point, as shown in Figure 5-1. This graphical correlation can be expressed mathematically by the following equation:

$$b = a_1 + a_2(T_b - 460) + a_3(T - 460)^2 + a_4(T_b - 460)^3 \\ + a_5/(T - 460)$$

(5-2)

where T_b = boiling point, °R

$a_1 - a_5$ = coefficients with the following values:

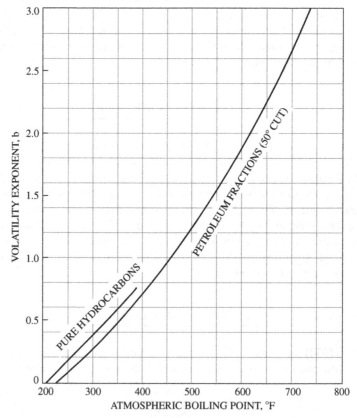

FIGURE 5-1   Volatility exponent. (Courtesy of the *Petroleum Refiner*.)

$a_1 = 1.6744337$
$a_2 = -3.4563079 \times 10^{-3}$
$a_3 = 6.1764103 \times 10^{-6}$
$a_4 = 2.4406839 \times 10^{-6}$
$a_5 = 2.9289623 \times 10^{2}$

5.3 KATZ'S METHOD

Katz et al. (1957) [25] suggested that a factor of 0.15 times the equilibrium ratio for the heptane component would give a reasonably close approximation to the equilibrium ratio for heptanes and heavier. This suggestion is expressed mathematically by the following equation:

$$K_{C_{7+}} = 0.15 K_{C_7} \qquad (5\text{-}3)$$

6

Applications of the Equilibrium Ratio in Reservoir Engineering

The vast amount of experimental and theoretical work that has been performed on equilibrium ratio studies indicates their importance in solving phase equilibrium problems in reservoir and process engineering. Some of their practical applications are discussed next.

6.1 DEW-POINT PRESSURE

The dew-point pressure p_d of a hydrocarbon system is defined as the pressure at which an infinitesimal quantity of liquid is in equilibrium with a large quantity of gas. For a total of 1 lb-mol of a hydrocarbon mixture, i.e., $n = 1$, the following conditions are applied at the dew-point pressure:

$$n_L = 0$$
$$n_v = 1$$

Under these conditions, the composition of the vapor phase y_i is equal to the overall composition z_i. Applying these constraints to Equation 2-14 yields

$$\sum_i \frac{z_i}{K_i} \qquad (6\text{-}1)$$

where z_i = total composition of the system under consideration.

The solution of Equation 6-1 for the dew-point pressure p_d involves a trial-and-error process. The process is summarized in the following steps:

Step 1. Assume a trial value of p_d. A good starting value can be obtained by applying Wilson's equation (Equation 4-1) for calculating K_i to Equation 6-1 to give

$$\sum_i \left\{ \frac{z_i}{\dfrac{P_{ci}}{P_d} \exp\left[5.37(1 + \omega_i)\left(1 - \dfrac{T_{ci}}{T}\right) \right]} \right\} = 1$$

Solving for p_d yields

$$\text{initial } p_d = \frac{1}{\sum_i \left\{ \dfrac{z_i}{P_{ci} \exp\left[5.37(1 + \omega_i)\left(1 - \dfrac{T_{ci}}{T}\right) \right]} \right\}} \qquad (6\text{-}2)$$

Another simplified approach for estimating the dew-point pressure is to treat the hydrocarbon mixture as an ideal system with the equilibrium ratio K_i as given by Equation 2-4:

$$K_i = \frac{P_{vi}}{P}$$

Substituting the above expression into Equation 5-1 gives

$$\sum_i \left[z_i \left(\frac{P_d}{P_{vi}} \right) \right] = 1.0$$

Solving for p_d yields

$$\text{initial } p_d = \frac{1}{\sum\limits_{i=1} \left(\dfrac{z_i}{P_{vi}} \right)}$$

Step 2. Using the assumed dew-point pressure, calculate the equilibrium ratio, K_i, for each component at the system temperature.

Step 3. Compute the summation of Equation 6-2.

Step 4. If the sum is less than 1, steps 2 and 3 are repeated at a higher initial value of pressure; conversely, if the sum is greater than 1, repeat the calculations with a lower initial value of p_d. The correct value of the dew-point pressure is obtained when the sum is equal to 1.

Example 6-1

A natural gas reservoir at 250°F has the following composition:

Component	z_i
C_1	0.80
C_2	0.05
C_3	0.04
$i - C_4$	0.03
$n - C_4$	0.02
$i - C_5$	0.03
$n - C_5$	0.02
C_6	0.005
C_{7+}	0.005

If the molecular weight and specific gravity of C_{7+} are 140 and 0.8, calculate the dew-point pressure.

Solution

Step 1. Calculate the convergence pressure of the mixture from Rzasa's correlation, i.e., Equation 4-11, to give

$$p_k = 5000 \text{ psia}$$

Step 2. Determine an initial value for the dew-point pressure from Equation 6-2 to give

$$p_d = 207 \text{ psia}$$

Step 3. Using the K-value curves in the Gas Processors Suppliers Association *Engineering Data Book*, solve for the dew-point pressure by applying the iterative procedure outlined previously, and using Equation 6-1, to give

Component	z_i	K_i at 207 psia	z_i/K_i	K_i at 300 psia	z_i/K_i	K_i at 222.3 psia	z_i/K_i
C_1	0.78	19	0.0411	13	0.06	18	0.0433
C_2	0.05	6	0.0083	4.4	0.0114	5.79	0.0086
C_3	0.04	3	0.0133	2.2	0.0182	2.85	0.0140

$i - C_4$	0.03	1.8	0.0167	1.35	0.0222	1.75	0.0171
$n - C_4$	0.02	1.45	0.0138	1.14	0.0175	1.4	0.0143
$i - C_5$	0.03	0.8	0.0375	0.64	0.0469	0.79	0.0380
$n - C_5$	0.02	0.72	0.0278	.55	0.0364	0.69	0.029
C_6	0.005	0.35	0.0143	0.275	0.0182	0.335	0.0149
C_{7+}	0.02	0.255*	0.7843	0.02025*	0.9877	0.0243*	0.8230
			0.9571		1.2185		1.0022

*Equation 5-1.

The dew-point pressure is therefore 222 psia at 250°F.

6.2 BUBBLE-POINT PRESSURE

At the bubble point, p_b, the hydrocarbon system is essentially liquid, except for an infinitesimal amount of vapor. For a total of 1 lb-mol of the hydrocarbon mixture, the following conditions are applied at the bubble-point pressure:

$$n_L = 1$$
$$n_v = 0$$

Obviously, under these conditions, $x_i = z_i$. Applying the preceding constraints to Equation 2-15 yields

$$\sum_i (z_i K_i) = 1 \tag{6-3}$$

Following the procedure outlined in the dew-point pressure determination, Equation 6-3 is solved for the bubble-point pressure p_b by assuming various pressures and determining the pressure that will produce K values that satisfy Equation 6-3.

During the iterative process, if

$$\sum_i (z_i K_i) < 1, \rightarrow \text{ the assumed pressure is high}$$

$$\sum_i (z_i K_i) > 1, \rightarrow \text{ the assumed pressure is low}$$

Wilson's equation can be used to give a good starting value for the iterative process:

$$\sum_i \left\{ z_i \frac{P_{ci}}{P_b} \exp\left[5.37(1 + \omega)\left(1 - \frac{T_{ci}}{T}\right)\right] \right\} = 1$$

Solving for the bubble-point pressure gives

$$P_b = \sum_i \left\{ z_i P_{ci} \exp\left[5.37(1 + \omega)\left(1 - \frac{T_{ci}}{T}\right)\right] \right\} \tag{6-4}$$

Assuming an ideal solution behavior, an initial guess for the bubble-point pressure can also be calculated by replacing the K_i in Equation 6-3 with that of Equation 2-4 to give

$$\sum_i \left[z_i \left(\frac{p_{vi}}{p_b} \right) \right] = 1$$

or

$$p_b = \sum_i (z_i p_{vi}) \qquad (6\text{-}5)$$

Example 6-2

A crude oil reservoir has a temperature of $200°F$ and a composition as follows. Calculate the bubble-point pressure of the oil.

Component	x_i
C_1	0.42
C_2	0.05
C_3	0.05
$i - C_4$	0.03
$n - C_4$	0.02
$i - C_5$	0.01
$n - C_5$	0.01
C_6	0.01
C_{7+}	0.40*

*$(M)_{C_{7+}} = 216.0$.
$(\gamma)_{C_{7+}} = 0.8605$.
$(T_b)_{C_{7+}} = 977°R$.

Solution

Step 1. Calculate the convergence pressure of the system by using Standing's correlation (Equation 4-10):

$$p_k = (60)(216) - 4200 = 8760 \text{ psia}$$

Step 2. Calculate the critical pressure and temperature using the equation of Riazi and Daubert (1987):

$$\theta = a(M)^b \gamma^c \exp[d(M) + e\gamma + f(M)\gamma]$$

where θ = any physical property
    a–f = constants for each property, as given in Table 6-1
    γ = specific gravity of the fraction
    M = molecular weight
    T_c = critical temperature, °R
    P_c = critical pressure, psia, as given in Table 6-1
    V_c = critical volume, ft^3/lb
    T_b = boiling-point temperature, °R
This gives

$$p_c = 230.4 \text{ psia}$$

$$T_c = 1279.8°R$$

Step 3. Calculate the acentric factor by employing the correlation of Edmister (1958) [12] for the acentric factor of pure fluids and petroleum fractions:

$$\omega = \frac{3[\log(p_c/14.70)]}{7[(T_c/T_b - 1)]}$$

where T = acentric factor
    p_c = critical pressure, psia
    T_c = critical temperature, °R
    T_b = normal boiling point, °R

This yields

$$\omega = 0.653$$

Step 4. Estimate the bubble-point pressure from Equation 6-4 to give

$$p_b = 3924 \text{ psia}$$

TABLE 6-1 Correlation Constants for the Riazi and Daubert Equation

θ	a	b	c	d	e	f
T_c, °R	544.4	0.2998	1.0555	-1.3478×10^{-4}	-0.61641	0.0
P_c, psia	4.5203×10^4	-0.8063	1.6015	-1.8078×10^{-3}	-0.3084	0.0
V_c, ft^3/lb	1.206×10^{-2}	0.20378	-1.3036	-2.657×10^{-3}	0.5287	2.6012×10^{-3}
T_b, °R	6.77857	0.401673	-1.58262	3.77409×10^{-3}	2.984036	-4.25288×10^{-3}

Step 5. Employing the iterative procedure outlined previously and using the Whitson and Torp equilibrium ratio correlation gives

Component	z_i	K_i at 3924 psia	$z_i K_i$	K_i at 3950 psia	$z_i K_i$	K_i at 4329 psia	$z_i K_i$
C_1	0.42	2.257	0.9479	2.242	0.9416	2.0430	0.8581
C_2	0.05	1.241	0.06205	2.137	0.0619	1.1910	0.0596
C_3	0.05	0.790	0.0395	0.7903	0.0395	0.793	0.0397
$i - C_4$	0.03	0.5774	0.0173	0.5786	0.0174	0.5977	0.0179
$n - C_4$	0.02	0.521	0.0104	0.5221	0.0104	0.5445	0.0109
$i - C_5$	0.01	0.3884	0.0039	0.3902	0.0039	0.418	0.0042
$n - C_5$	0.01	0.3575	0.0036	0.3593	0.0036	0.3878	0.0039
C_6	0.01	0.2530	0.0025	0.2549	0.0025	0.2840	0.0028
C_{7+}	0.40	0.227	0.0091	0.0232	0.00928	0.032	0.0138
Σ			1.09625		1.09008		1.0099

The calculated bubble-point pressure is 4330 psia.

6.3 SEPARATOR CALCULATIONS

Produced reservoir fluids are complex mixtures of different physical characteristics. As a well stream flows from the high-temperature, high-pressure petroleum reservoir, it experiences pressure and temperature reductions. Gases evolve from the liquids and the well stream changes in character. The physical separation of these phases is by far the most common of all field-processing operations and one of the most critical. The manner in which the hydrocarbon phases are separated at the surface influences the stock-tank oil recovery. The principal means of surface separation of gas and oil is the conventional stage separation.

Stage separation is a process in which gaseous and liquid hydrocarbons are flashed (separated) into vapor and liquid phases by two or more separators. These separators are usually operated in series at consecutively lower pressures. Each condition of pressure and temperature at which hydrocarbon phases are flashed is called a *stage of separation.* Examples of one- and two-stage separation processes are shown in Figure 6-1. Traditionally, the stock tank is normally considered a separate stage of separation. Mechanically, there are two types of gas-oil separation: (1) differential separation and (2) flash or equilibrium separation.

To explain the various separation processes, it is convenient to define the composition of a hydrocarbon mixture by three groups of components:

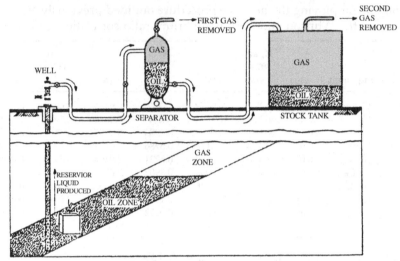

One-Stage Separation

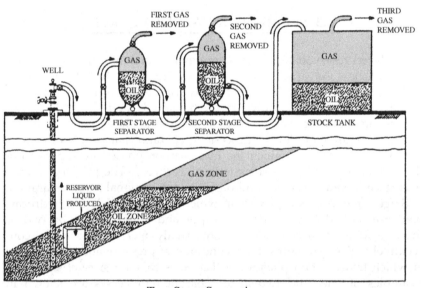

Two-Stage Separation

FIGURE 6-1   Schematic drawing of one- and two-stage separation processes. (After Clark, 1951.)

1. The very volatile components ("lights"), such as nitrogen, methane, and ethane.
2. The components of intermediate volatility ("intermediates"), such as propane through hexane.
3. The components of less volatility, or the "heavies," such as heptane and heavier components.

In differential separation, the liberated gas (which is composed mainly of lighter components) is removed from contact with the oil as the pressure on the oil is reduced. As pointed out by Clark (1960) [10], when the gas is separated in this manner, the maximum amount of heavy and intermediate components will remain in the liquid, minimum shrinkage of the oil will occur, and, therefore, greater stock-tank oil recovery will occur. This is due to the fact that the gas liberated earlier at higher pressures is not present at lower pressures to attract the intermediate and heavy components and pull them into the gas phase.

In flash (equilibrium) separation, the liberated gas remains in contact with the oil until its instantaneous removal at the final separation pressure. A maximum proportion of intermediate and heavy components is attracted into the gas phase by this process, and this results in a maximum oil shrinkage and, thus, a lower oil recovery.

In practice, the differential process is introduced first in field separation when gas or liquid is removed from the primary separator. In each subsequent stage of separation, the liquid initially undergoes a flash liberation followed by a differential process as actual separation occurs. As the number of stages increases, the differential aspect of the overall separation becomes greater.

The purpose of stage separation, then, is to reduce the pressure on the produced oil in steps so that more stock-tank oil recovery will result. Separator calculations are basically performed to determine

- Optimum separation conditions: separator pressure and temperature.
- Compositions of the separated gas and oil phases.
- Oil formation volume factor.
- Producing gas-oil ratio.
- API gravity of the stock-tank oil.

Note that, if the separator pressure is high, large amounts of light components will remain in the liquid phase at the separator and be lost along with other valuable components to the gas phase at the stock tank. On the other hand, if the pressure is too low, large amounts of light components will be separated from the liquid and they will attract substantial quantities of intermediate and heavier components. An intermediate pressure, called optimum separator pressure, should be selected to maximize the oil volume accumulation in the stock tank. This optimum pressure will also yield

- A maximum stock-tank API gravity.
- A minimum oil formation volume factor (i.e., less oil shrinkage).
- A minimum producing gas-oil ratio (gas solubility).

The concept of determining the optimum separator pressure by calculating the API gravity, B_o, and R_s is shown graphically in Figure 6-2. The computational steps of the separator calculations are described in

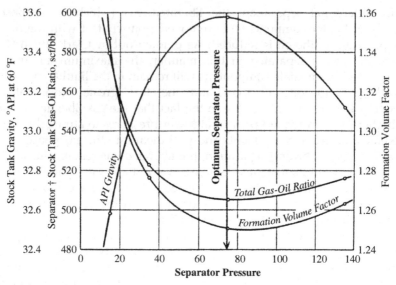

FIGURE 6-2   Effect of separator pressure on API, B_o, and GOR. (After Amyx, Bass, and Whitney, 1960.)

conjunction with Figure 6-3, which schematically shows a bubble-point reservoir flowing into a surface separation unit consisting of n stages operating at successively lower pressures.

Step 1. Calculate the volume of oil occupied by 1 lb-mol of crude at the reservoir pressure and temperature. This volume, denoted V_o, is calculated by recalling and applying the equation that defines the number of moles to give

$$n = \frac{m}{M_a} = \frac{\rho_o V_o}{M_a} = 1$$

Solving for the oil volume gives

$$V_o = \frac{M_a}{\rho_o} \tag{6-6}$$

where m = total weight of 1 lb-mol of crude oil, lb/mol
  V_o = volume of 1 lb-mol of crude oil at reservoir conditions, ft³/mol
  M_a = apparent molecular weight
  ρ_o = density of the reservoir oil, lb/ft³

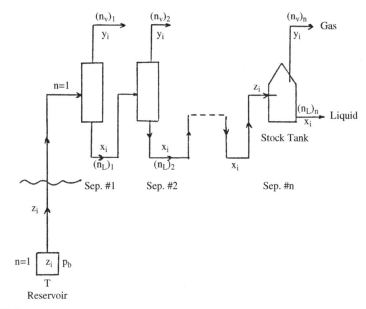

FIGURE 6-3 Schematic illustration of n separation stages.

Step 2. Given the composition of the feed stream z_i to the first separator and the operating conditions of the separator, i.e., separator pressure and temperature, calculate the equilibrium ratios of the hydrocarbon mixture.

Step 3. Assuming a total of 1 mol of the feed entering the first separator and using the previously calculated equilibrium ratios, perform flash calculations to obtain the compositions and quantities, in moles, of the gas and the liquid leaving the first separator. Designating these moles as $(n_L)_1$ and $(n_v)_1$, the actual numbers of moles of the gas and the liquid leaving the first separation stage are

$$[n_{v1}]_a = (n)(n_v)_1 = (1)(n_v)_1$$

$$[n_{L1}]_a = (n)(n_L)_1 = (1)(n_L)_1$$

where $[n_{v1}]_a$ = actual number of moles of vapor leaving the first separator

$\quad\quad [n_{L1}]_a$ = actual number of moles of liquid leaving the first separator

Step 4. Using the composition of the liquid leaving the first separator as the feed for the second separator, i.e., $z_i = x_i$, calculate the equilibrium ratios of the hydrocarbon mixture at the prevailing pressure and temperature of the separator.

Step 5. Based on 1 mol of the feed, perform flash calculations to determine the compositions and quantities of the gas and liquid leaving the second separation stage. The actual numbers of moles of the two phases are then calculated from

$$[n_{v2}]_a = [n_{L1}]_a(n_v)_2 = (1)(n_L)_1(n_v)_2$$
$$[n_{L2}]_a = [n_{L1}]_a(n_L)_2 = (1)(n_L)_1(n_L)_2$$

where $[n_{v2}]_a$, $[n_{L2}]_a$ = actual moles of gas and liquid leaving
    separator 2
  $(n_v)_2$, $(n_L)_2$ = moles of gas and liquid as determined
    from flash calculations

Step 6. The previously outlined procedure is repeated for each separation stage, including the stock-tank storage, and the calculated moles and compositions are recorded. The total number of moles of gas off all stages is then calculated as

$$(n_v)_t = \sum_{i=1}^{n}(n_{va})_i = (n_v)_1 + (n_L)_1(n_v)_2 + (n_L)_1(n_L)_2(n_v)_3$$
$$+ \ldots + (n_L)_1 \ldots (n_L)_{n-1}(n_v)_n$$

In a more compacted form, this expression can be written

$$(n_v)_t = (n_v)_1 + \sum_{i=2}^{n}\left[(n_v)_i\prod_{j=1}^{i-1}(n_L)_j\right] \tag{6-7}$$

where $(n_v)_t$ = total moles of gas off all stages, lb-mol/mol
    of feed
  n = number of separation stages

Total moles of liquid remaining in the stock tank can also be calculated as

$$(n_L)_{st} = n_{L1}n_{L2}\ldots n_{Ln}$$

or

$$(n_L)_{st} = \prod_{i=1}^{n}(n_L)_i \tag{6-8}$$

where $(n_L)_{st}$ = moles of liquid remaining in the stock tank
  $(n_L)_i$ = moles of liquid off ith stage.

Step 7. Calculate the volume, in scf, of all the liberated solution gas from

$$V_g = 379.4(n_v)_t \tag{6-9}$$

where V_g = total volume of the liberated solution gas, scf/mol
    of feed.

Step 8. Determine the volume of stock-tank oil occupied by $(n_L)_{st}$ moles of liquid from

$$(V_o)_{st} = \frac{(n_L)_{st}(M_a)_{st}}{(\rho_o)_{st}} \tag{6-10}$$

where $(V_o)_{st}$ = volume of stock-tank oil, ft^3/mol of feed
$(M_a)_{st}$ = apparent molecular weight of the stock-tank oil
$(\rho_o)_{st}$ = density of the stock-tank oil, lb/ft^3

Step 9. Calculate the specific gravity and the API gravity of the stock-tank oil by applying these expressions:

$$\gamma_o = \frac{(\rho_o)_{st}}{62.4}$$

$$°API = \frac{141.5}{\gamma_o} - 131.5$$

Step 10. Calculate the total gas-oil ratio (or gas solubility R_s):

$$GOR = \frac{V_g}{(V_o)_{st}/5.615} = \frac{(5.615)(379.4)(n_v)_t}{(n_L)_{st}(M)_{st}/(\rho_o)_{st}}$$

$$GOR = \frac{2,130.331(n_v)_t(\rho_o)_{st}}{(n_L)_{st}(M)_{st}} \tag{6-11}$$

where GOR = gas-oil ratio, scf/STB.

Step 11. Calculate the oil formation volume factor from the relationship

$$B_o = \frac{V_o}{(V_o)_{st}}$$

Combining Equations 6-6 and 6-10 with the preceding expression gives

$$B_o = \frac{M_a(\rho_o)_{st}}{\rho_o(n_L)_{st}(M_a)_{st}} \tag{6-12}$$

where B_o = oil formation volume factor, bbl/STB
M_a = apparent molecular weight of the feed
$(M_a)_{st}$ = apparent molecular weight of the stock-tank oil
ρ_o = density of crude oil at reservoir conditions, lb/ft^3

The separator pressure can be optimized by calculating the API gravity, GOR, and B_o in the manner just outlined at different assumed pressures. The optimum pressure corresponds to a maximum in the API gravity and a minimum in the gas-oil ratio and oil formation volume factor.

Example 6-3

A crude oil, with the following composition, exists at its bubble-point pressure of 1708.7 psia and at a temperature of 131°F. The crude oil is flashed through two-stage and stock-tank separation facilities. The operating conditions of the three separators are

Separator	Pressure, psia	Temperature, °F
1	400	72
2	350	72
Stock tank	14.7	60

The composition of the crude oil follows:

Component	z_i
CO_2	0.0008
N_2	0.0164
C_1	0.2840
C_2	0.0716
C_3	0.1048
$i - C_4$	0.0420
$n - C_4$	0.0420
$i - C_5$	0.0191
$n - C_5$	0.0191
C_6	0.0405
C_{7+}	0.3597

The molecular weight and specific gravity of C_{7+} are 252 and 0.8429. Calculate B_o, R_s, stock-tank density, and the API gravity of the hydrocarbon system.

Solution

Step 1. Calculate the apparent molecular weight of the crude oil to give $M_a = 113.5102$.

Step 2. Calculate the density of the bubble-point crude oil by using the Standing and Katz correlation to yield $\rho_o = 44.794 \text{ lb/ft}^3$.

Step 3. Flash the original composition through the first separator by generating the equilibrium ratios using the Standing correlation (Equation 4-4) to give

Component	z_i	K_i	x_i	y_i
CO_2	0.0008	3.509	0.0005	0.0018
N_2	0.0164	39.90	0.0014	0.0552
C_1	0.2840	8.850	0.089	0.7877

Component	z_i	K_i	x_i	y_i
C_2	0.0716	1.349	0.0652	0.0880
C_3	0.1048	0.373	0.1270	0.0474
$i - C_4$	0.0420	0.161	0.0548	0.0088
$n - C_4$	0.0420	0.120	0.0557	0.0067
$i - C_5$	0.0191	0.054	0.0259	0.0014
$n - C_5$	0.0191	0.043	0.0261	0.0011
C_6	0.0405	0.018	0.0558	0.0010
C_{7+}	0.3597	0.0021	0.4986	0.0009

with $n_L = 0.7209$ and $n_v = 0.29791$.

Step 4. Use the calculated liquid composition as the feed for the second separator and flash the composition at the operating condition of the separator:

Component	z_i	K_i	x_i	y_i
CO_2	0.0005	3.944	0.0005	0.0018
N_2	0.0014	46.18	0.0008	0.0382
C_1	0.089	10.06	0.0786	0.7877
C_2	0.0652	1.499	0.0648	0.0971
C_3	0.1270	0.4082	0.1282	0.0523
$i - C_4$	0.0548	0.1744	0.0555	0.0097
$n - C_4$	0.0557	0.1291	0.0564	0.0072
$i - C_5$	0.0259	0.0581	0.0263	0.0015
$n - C_5$	0.0261	0.0456	0.0264	0.0012
C_6	0.0558	0.0194	0.0566	0.0011
C_{7+}	0.4986	0.00228	0.5061	0.0012

with $n_L = 0.9851$ and $n_v = 0.0149$.

Step 5. Repeat the above calculation for the stock-tank stage to give

Component	z_i	K_i	x_i	y_i
CO_2	0.0005	81.14	0000	0.0014
N_2	0.0008	1159	0000	0.026
C_1	0.0784	229	0.0011	0.2455
C_2	0.0648	27.47	0.0069	0.1898
C_3	0.1282	6.411	0.0473	0.3030
$i - C_4$	0.0555	2.518	0.0375	0.0945
$n - C_4$	0.0564	1.805	0.0450	0.0812
$i - C_5$	0.0263	0.7504	0.0286	0.0214
$n - C_5$	0.0264	0.573	0.02306	0.0175
C_6	0.0566	0.2238	0.0750	0.0168
C_{7+}	0.5061	0.03613	0.7281	0.0263

with $n_L = 0.6837$ and $n_v = 0.3163$.

Step 6. Calculate the actual number of moles of the liquid phase at the stock-tank conditions from Equation 6-8:

$$(n_L)_{st} = (1)(0.7209)(0.9851)(0.6837) = 0.48554$$

Step 7. Calculate the total number of moles of the liberated gas from the entire surface separation system:

$$n_v = 1 - (n_L)_{st} = 1 - 0.48554 = 0.51446$$

Step 8. Calculate apparent molecular weight of the stock-tank oil from its composition to give $(M_a)_{st} = 200.6$.

Step 9. Calculate the density of the stock-tank oil using the Standing correlation to give

$$(\rho_o)_{st} = 50.920$$
$$\gamma = 50.920/62.4 = 0.816 \ 60°/60°$$

Step 10. Calculate the API gravity of the stock-tank oil:

$$API = (141.5/0.816) - 131.5 = 41.9$$

Step 11. Calculate the gas solubility from Equation 6-11 to give

$$R_s = \frac{2130.331(0.51446)(50.92)}{0.48554(200.6)} = 573.0 \ scf/STB$$

Step 12. Calculate B_o from Equation 6-12 to give

$$B_o = \frac{(113.5102)(50.92)}{(44.794)(0.48554)(200.6)} = 1.325 \ bbl/STB$$

To optimize the operating pressure of the separator, these steps should be repeated several times under different assumed pressures, and the results, in terms of API, B_o, and R_s, should be expressed graphically and used to determine the optimum pressure.

Note that, at *low pressures*, e.g., $p < 1000$, equilibrium ratios are nearly independent of the overall composition z_i or the convergence pressure and can be considered only a function of pressure and temperature. Under this condition, i.e, $p < 1000$, the equilibrium ratio for any component i can be expressed as

$$K_i = \frac{A_i}{p}$$

The temperature-dependent coefficient A_i is a characterization parameter of component i that accounts for the physical properties of the component. The preceding expression suggests that K_i varies linearly at a constant temperature with $1/p$. For example, suppose that a hydrocarbon mixture exists at 300 psi and 100°F. Assume that the mixture contains methane and we want to estimate the equilibrium ratio of methane (or any other components) when the mixture is flashed at 100 psi and at the same temperature of 100°F. The recommended procedure is summarized in the following steps:

Step 1. Because at low pressure the equilibrium ratio is considered independent of the overall composition of the mixture, use the equilibrium ratio charts of the Gas Processors Suppliers Association *Engineering Data Book* to determine the K_i value of methane at 300 psi and 100°F:

$$K_{C_1} = 10.5$$

Step 2. Calculate the characterization parameter A_i of methane from this proposed relationship:

$$10.5 = \frac{A_i}{500}$$

$$A_i = (10.5)(300) = 3{,}150$$

Step 3. Calculate the K_i of methane at 100 psi and 100°F from

$$K_{C_1} = \frac{3{,}150}{100} = 31.5$$

In many low-pressure applications of flash calculations at constant temperature, it might be possible to characterize the entire hydrocarbon mixture as a binary system, i.e., two-component system. Because methane exhibits a linear relationship with pressure of a wide range of pressure values, one of the components that form the binary system should be methane. The main advantage of such a binary system is the simplicity of performing flash calculations because it does not require an iterative technique.

Reconsider Example 6-3 where flash calculations were performed on the entire system at 400 psia and 72°F. To perform flash calculations on the feed for the second separator at 350 psi and 72°F, follow these steps:

Step 1. Select methane as one of the binary systems with the other component defined as ethane-plus, i.e., C_{2+}, which lumps the remaining components. The results of Example 6-3 show

- $K_{C_1} = 8.85$.
- $y_{C_1} = 0.7877$.
- $x_{C_2} = 0.089$.
- $y_{C_{2+}} = 1.0 - 0.7877 = 0.2123$.
- $x_{C_{2+}} = 1.0 - 0.089 = 0.911$.

Step 2. From the definition of the equilibrium ratio, calculate the K value of C_{2+}:

$$K_{C_{2+}} = \frac{y_{C_{2+}}}{x_{C_{2+}}} = \frac{0.2123}{0.9110} = 0.2330$$

Step 3. Calculate the characterization parameter A_i for methane and C_{2+}:

$$A_{C_1} = K_{C_1}p = (8.85)(400) = 3,540$$
$$A_{C_{2+}} = K_{C_{2+}}p = (0.233)(400) = 93.2$$

The equilibrium ratio for each of the two components (at a constant temperature) can then be described by

$$K_{C_1} = \frac{3,540}{p}$$
$$K_{C_{2+}} = \frac{93.2}{p}$$

Step 4. Calculate the K_i value for each component at the second separator pressure of 350 psi:

$$K_{C_1} = \frac{3,540}{350} = 10.11$$
$$K_{C_{2+}} = \frac{93.2}{350} = 0.266$$

Step 5. Using the flash calculations procedure as outlined previously for a binary system, calculate the composition and number of moles of the gas and liquid phase at 350 psi.

- Solve for x_{C1} and x_{C2+}:

$$x_{C_1} = \frac{1 - K_2}{K_1 - K_2} = \frac{1.0 - 0.266}{10.11 - 0.266} = 0.0746$$
$$x_{C_{2+}} = 1 - x_{C_1} = 1.0 - 0.0746 = 0.9254$$

- Solve for y_{C1} and y_{C2+}:

$$y_{C_1} = x_{C_1} K_1 = (0.0746)(10.11) = 0.754$$
$$y_{C_{2+}} = 1 - y_{C_1} = 1.0 - 0.754 = 0.246$$

- Solve for number of moles of the vapor and liquid phase:

$$n_v = \frac{z_1 - x_1}{x_1(K_1 - 1)} = \frac{0.089 - 0.0746}{0.0746(10.11 - 1)} = 0.212$$
$$n_L = 1 - n_v = 1.0 - 0.212 = 0.788$$

These calculations are considered meaningless without converting moles of liquid n_l into volume, which requires the calculation of the liquid density at separator pressure and temperature. Notice that

$$V = \frac{n_L M_a}{\rho_o}$$

where M_a is the apparent molecular weight of the separated liquid and is given by (for a binary system)

$$M_a = x_{C_1} M_{C_1} + x_{C_{2+}} M_{C_{2+}}$$

6.4 DENSITY CALCULATIONS

The calculation of crude oil density from its composition is an important and integral part of performing flash calculations. The best known and most widely used calculation methods are those of Standing and Katz (1942) [58] and Alani and Kennedy (1960) [3]. These two methods are presented next.

The Standing-Katz Method

Standing and Katz (1942) [58] proposed a graphical correlation for determining the density of hydrocarbon liquid mixtures. The authors developed the correlation from evaluating experimental, compositional, and density data on 15 crude oil samples containing up to 60 mol% methane. The proposed method yielded an average error of 1.2% and a maximum error of 4% for the data on these crude oils. The original correlation did not have a procedure for handling significant amounts of nonhydrocarbons.

The authors expressed the density of hydrocarbon liquid mixtures as a function of pressure and temperature by the following relationship:

$$\rho_o = \rho_{sc} + \Delta\rho_p - \Delta\rho_T$$

where ρ_o = crude oil density at p and T, lb/ft^3
ρ_{sc} = crude oil density (with all the dissolved solution gas) at standard conditions, i.e., 14.7 psia and 60°F, lb/ft^3
$\Delta\rho_p$ = density correction for compressibility of oils, lb/ft^3
$\Delta\rho_T$ = density correction for thermal expansion of oils, lb/ft^3

Standing and Katz correlated graphically the liquid density at standard conditions with

- The density of the propane-plus fraction $\rho_{C_{3+}}$.
- The weight percent of methane in the entire system $(m_{C_1})_{C_{1+}}$.
- The weight percent of ethane in the ethane-plus $(m_{C_2})_{C_{2+}}$.

This graphical correlation is shown in Figure 6-4. The following are the specific steps in the Standing and Katz procedure for calculating the liquid density at a specified pressure and temperature.

Step 1. Calculate the total weight and the weight of each component in 1 lb-mol of the hydrocarbon mixture by applying the following relationships:

$$m_i = x_i M_i$$
$$m_t = \Sigma x_i M_i$$

where m_i = weight of component i in the mixture, lb/lb-mol
x_i = mole fraction of component i in the mixture
M_i = molecular weight of component i
m_t = total weight of 1 lb-mol of the mixture, lb/lb-mol

Step 2. Calculate the weight percent of methane in the entire system and the weight percent of ethane in the ethane-plus from the following expressions:

$$(m_{C_1})_{C_{1+}} = \left[\frac{x_{C_1} M_{C_1}}{\sum\limits_{i=1}^{n} x_i M_i}\right] 100 = \left[\frac{m_{C_1}}{m_t}\right] 100$$

and

$$(m_{C_2})_{C_{2+}} = \left[\frac{m_{C_2}}{m_{C_{2+}}}\right] 100 = \left[\frac{m_{C_2}}{m_t - m_{C_1}}\right] 100$$

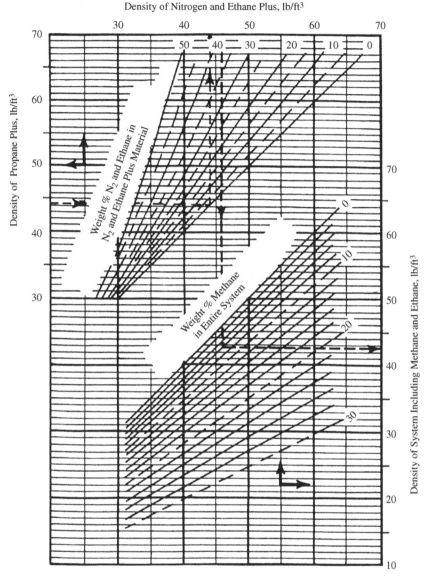

FIGURE 6-4 Standing and Katz density correlation. (Courtesy of the Gas Processors Suppliers Association, *Engineering Data Book*, 10th ed., 1987.)

where $(m_{C_1})_{C_{1+}}$ = weight percent of methane in the entire system

m_{C_1} = weight of methane in 1 lb-mol of the mixture, i.e., $x_{C_1} M_{C_1}$

$(m_{C_2})_{C_{2+}}$ = weight percent of ethane in ethane-plus

m_{C_2} = weight of ethane in 1 lb-mol of the mixture, i.e., $x_{C_2} M_{C_2}$

M_{C_1} = molecular weight of methane

M_{C_2} = molecular weight of ethane

Step 3. Calculate the density of the propane-plus fraction at standard conditions using the following equations:

$$\rho_{C_{3+}} = \frac{m_{C_3}}{V_{C_{3+}}} = \frac{\displaystyle\sum_{i=C_3}^{n} x_i M_i}{\displaystyle\sum_{i=C_3}^{n} \frac{x_i M_i}{\rho_{oi}}}$$

with

$$m_{C_{3+}} = \sum_{i=C_3} x_i M_i$$

$$V_{C_{3+}} = \sum_{i=C_3} V_i = \sum_{i=C_3} \frac{m_i}{\rho_{oi}}$$

where $\rho_{C_{3+}}$ = density of the propane and heavier components, lb/ft^3

$m_{C_{3+}}$ = weight of the propane and heavier fractions, $lb/lb\text{-mol}$

$V_{C_{3+}}$ = volume of the propane-plus fraction, $ft^3/lb\text{-mol}$

V_i = volume of component i in 1 lb-mol of the mixture

m_i = weight of component i, i.e., $x_i M_i$, $lb/lb\text{-mol}$

ρ_{oi} = density of component i at standard conditions, lb/ft^3

The density of the plus fraction must be measured.

Step 4. Using Figure 6-4, enter the $\rho_{C_{3+}}$ value into the left ordinate of the chart and move horizontally to the line representing $(m_{C_2})_{C_{2+}}$; then drop vertically to the line representing $(m_{C_1})_{C_{1+}}$. The density of the oil at standard condition is read on the right side of the chart. Standing (1977) [56] expressed the graphical correlation in the following mathematical form:

$$\rho_{sc} = \rho_{C_{2+}} \left[1 - 0.012(m_{C_1})_{C_{1+}} - 0.000158(m_{C_1})_{C_{1+}}^2 \right] \\ + 0.0133(m_{C_1})_{C_{1+}} + 0.00058(m_{C_1})_{C_{2+}}^2$$

with

$$\rho_{C_{2+}} = \rho_{C_{3+}} \left[1 - 0.01386(m_{C_2})_{C_{2+}} - 0.000082(m_{C_2})_{C_{2+}}^2 \right] \\ + 0.379(m_{C_2})_{C_{2+}} + 0.0042(m_{C_2})_{C_{2+}}^2$$

where $\rho_{C_{2+}}$ = density of ethane-plus fraction.

Step 5. Correct the density at standard conditions to the actual pressure by reading the additive pressure correction factor, $\Delta\rho_p$, from Figure 6-5, or using the following expression:

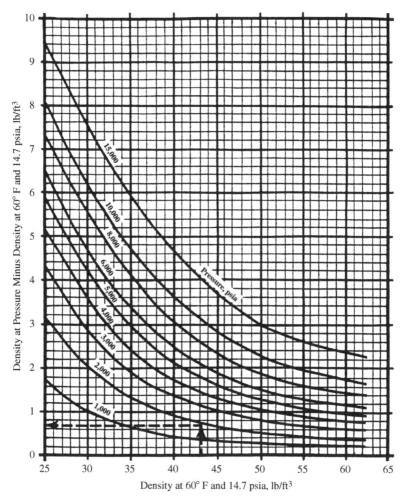

FIGURE 6-5 Density correction for compressibility of crude oils. (Courtesy of the Gas Processors Suppliers Association, *Engineering Data Book*, 10th ed., 1987.)

$$\Delta\rho_p = \left[0.000167 + (0.016181)10^{-0.0425\rho_{sc}}\right] p$$
$$- (10^{-8}) \left[0.299 + (263)10^{-0.0603\rho_{sc}}\right] p^2$$

Step 6. Correct the density at 60°F and pressure to the actual temperature by reading the thermal expansion correction term, $\Delta\rho_T$, from Figure 6-6, or from

$$\Delta\rho_T = (T - 520)\left[0.0133 + 152.4\left(\rho_{sc} + \Delta\rho_p\right)^{-2.45}\right]$$
$$- (T - 520)^2\left[8.1(10^{-6}) - (0.0622)10^{-0.0764(\rho_{sc}+\Delta\rho_p)}\right]$$

where T is the system temperature in °R.

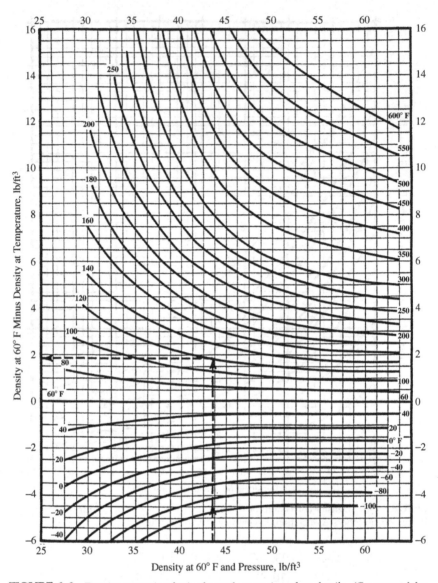

FIGURE 6-6   Density correction for isothermal expansion of crude oils. (Courtesy of the Gas Processors Suppliers Association, *Engineering Data Book*, 10th ed., 1987.)

Example 6-4

A crude oil system has the following composition:

Component	x_i
C_1	0.45
C_2	0.05
C_3	0.05
C_4	0.03
C_5	0.01
C_6	0.01
C_{7+}	0.40

If the molecular weight and specific gravity of C_{7+} fractions are 215 and 0.87, respectively, calculate the density of the crude oil at 4000 psia and 160°F using the Standing and Katz method.

Solution

Component	x_i	M_i	$m_i = x_i M_i$	ρ_{oi}, lb/ft³	$V_i = m_i/\rho_{oi}$
C_1	0.45	16.04	7.218	—	—
C_2	0.05	30.07	1.5035	—	—
C_3	0.05	44.09	2.2045	31.64	0.0697
C_4	0.03	58.12	1.7436	35.71	0.0488
C_5	0.01	72.15	0.7215	39.08	0.0185
C_6	0.01	86.17	0.8617	41.36	0.0208
C_{7+}	0.40	215.0	86.00	54.288*	1.586
			$m_t = 100.253$		$V_{C_{3+}} = 1.7418$

$*\rho_{C_{7+}} = (0.87)(62.4) = 54.288.$

Step 1. Calculate the weight percent of C_1 in the entire system and the weight percent of C_2 in the ethane-plus fraction:

$$(m_{C_1})_{C_{1+}} = \left[\frac{7.218}{100.253} \right] 100 = 7.2\%$$

$$(m_{C_2})_{C_{2+}} = \left[\frac{1.5035}{100.253 - 7.218} \right] 100 = 1.616\%$$

Step 2. Calculate the density of the propane-plus fraction:

$$\rho_{C_{3+}} = \frac{100.253 - 7.218 - 1.5035}{1.7418} = 52.55 \text{ lb/ft}^3$$

Step 3. Determine the density of the oil at standard conditions from Figure 6-4:

$$\rho_{sc} = 47.5 \text{ lb/ft}^3$$

Step 4. Correct for the pressure using Figure 6-5:

$$\Delta\rho_p = 1.18 \text{ lb/ft}^3$$

Density of the oil at 4000 psia and 60°F is then calculated by the expression

$$\rho_{p,60} = \rho_{sc} + \Delta\rho_p = 47.5 + 1.18 = 48.68 \text{ lb/ft}^3$$

Step 5. From Figure 6-6, determine the thermal expansion correction factor:

$$\Delta\rho_T = 2.45 \text{ lb/ft}^3$$

Step 6. The required density at 4000 psia and 160°F is

$$\rho_0 = 48.68 - 2.45 = 46.23 \text{ lb/ft}^3$$

The Alani-Kennedy Method

Alani and Kennedy (1960) [3] developed an equation to determine the molar liquid volume V_m of pure hydrocarbons over a wide range of temperature and pressure. The equation was then adopted to apply to crude oils, with the heavy hydrocarbons expressed as a heptanes-plus fraction, i.e., C_{7+}.

The Alani-Kennedy equation is similar in form to the van der Waals equation, which takes the following form:

$$V_m^3 - \left[\frac{RT}{P} + b\right] V_m^2 + \frac{aV_m}{P} - \frac{ab}{P} = 0 \tag{6-13}$$

where R = gas constant, 10.73 psia ft³/lb-mol °R
       T = temperature, °R
       p = pressure, psia
       V_m = molecular volume, ft³/lb-mol
       a, b = constants for pure substances

Alani and Kennedy considered the constants a and b to be functions of temperature and proposed these expressions for calculating the two parameters:

$$a = Ke^{n/T}$$
$$b = mT + c$$

where K, n, m, and c are constants for each pure component. Values of these constants are tabulated in Table 6-2. Table 6-2 contains no constants from which the values of the parameters a and b for heptanes-plus can be calculated. Therefore, Alani and Kennedy proposed the following equations for determining a and b of C_{7+}:

$$\ln(a_{C_{7+}}) = 3.8405985(10^{-3})(M)_{C_{7+}} - 9.5638281(10^{-4})\left(\frac{M}{\gamma}\right)_{C_{7+}}$$

$$+ \frac{261.80818}{T} + 7.3104464(10^{-6})(M)_{C_{7+}}^2 + 10.753517$$

$$b_{C_{7+}} = 0.03499274(M)_{C_{7+}} - 7.275403(\gamma)_{C_{7+}} + 2.232395(10^{-4})T$$

$$- 0.016322572\left(\frac{M}{\gamma}\right)_{C_{7+}} + 6.2256545$$

TABLE 6-2 Alani-Kennedy Coefficients

Components	K	n	m × 10⁴	c
C_1 70° – 300°F	9,160.6413	61.893223	3.3162472	0.50874303
C_1 301° – 460°F	147.47333	3,247.4533	−14.072637	1.8326659
C_2 100° – 249°F	46,709.573	−404.48844	5.1520981	0.52239654
C_2 250° – 460°F	17,495.343	34.163551	2.8201736	0.62309877
C_3	20,247.757	190.24420	2.1586448	0.90832519
$i - C_4$	32,204.420	131.63171	3.3862284	1.1013834
$n - C_4$	33,016.212	146.15445	2.902157	1.1168144
$i - C_5$	37,046.234	299.62630	2.1954785	1.4364289
$n - C_5$	37,046.234	299.62630	2.1954785	1.4364289
$n - C_6$	52,093.006	254.56097	3.6961858	1.5929406
H_2S*	13,200.00	0	17.900	0.3945
N_2*	4,300.00	2.293	4.490	0.3853
CO_2*	8,166.00	126.00	1.8180	0.3872

*Values for nonhydrocarbon components as proposed by Lohrenz et al. (1964) [22].

where $M_{C_{7+}}$ = molecular weight of C_{7+}
    $\gamma_{C_{7+}}$ = specific gravity of C_{7+}
    $a_{C_{7+}}, b_{C_{7+}}$ = constants of the heptanes-plus fraction
    T = temperature in °R

For hydrocarbon mixtures, the values of a and b of the mixture are calculated using the following mixing rules:

$$a_m = \sum_{i=1}^{C_{7+}} a_i x_i$$

$$b_m = \sum_{i=1}^{C_{7+}} b_i x_i$$

where the coefficients a_i and b_i refer to pure hydrocarbons at existing temperature and x_i is the mole fraction in the mixture. The values of a_m and b_m are then used in Equation 6-13 to solve for the molar volume V_m. The density of the mixture at pressure and temperature of interest is determined from the following relationship:

$$\rho_o = \frac{M_a}{V_m}$$

where ρ_o = density of the crude oil, lb/ft^3
    M_a = apparent molecular weight, i.e., $M_a = \sum x_i M_i$
    V_m = molar volume, ft^3/lb-mol

The Alani-Kennedy method for calculating the density of liquids is summarized in the following steps.

Step 1. Calculate the constants a and b for each pure component from

$$a = K e^{n/T}$$
$$b = mT + c$$

Step 2. Determine $a_{C_{7+}}$, and $b_{C_{7+}}$.
Step 3. Calculate the values of coefficients a_m and b_m.
Step 4. Calculate molar volume V_m by solving Equation 6-13 for the smallest real root:

$$V_m^3 - \left[\frac{RT}{P} + b_m\right] V_m^2 + \frac{a_m V_m}{P} - \frac{a_m b_m}{P} = 0$$

Step 5. Compute the apparent molecular weight, M_a.
Step 6. Determine the density of the crude oil from

$$\rho_0 = \frac{M_a}{V_m}$$

Example 6-5

A crude oil system has the following composition:

Component	x_i
CO_2	0.0008
N_2	0.0164
C_1	0.2840
C_2	0.0716
C_3	0.1048
$i - C_4$	0.0420
$n - C_4$	0.0420
$i - C_5$	0.0191
$n - C_5$	0.0191
C_6	0.0405
C_{7+}	0.3597

The following additional data are given:

$$M_{C_{7+}} = 252$$

$$\gamma_{C_{7+}} = 0.8424$$

$$\text{Pressure} = 1708.7 \text{ psia}$$
$$\text{Temperature} = 591°R$$

Calculate the density of the crude oil.

Solution

Step 1. Calculate the parameters $a_{C_{7+}}$ and $b_{C_{7+}}$:

$$a_{C_{7+}} = 229269.9$$
$$b_{C_{7+}} = 4.165811$$

Step 2. Calculate the mixture parameters a_m and b_m:

$$a_m = \sum_{i=1}^{C_{7+}} a_i x_i$$

$$a_m = 99111.71$$

$$b_m = \sum_{i=1}^{C_{7+}} b_i x_i$$

$$b_m = 2.119383$$

Step 3. Solve Equation 6-1 for the molar volume:

$$V_m^3 - \left[\frac{RT}{p} + b_m\right]V_m^2 + \frac{a_m V_m}{p} - \frac{a_m b_m}{p} = 0$$

$$V_m = 2.528417$$

Step 4. Determine the apparent molecular weight of this mixture:

$$M_a = \Sigma x_i M_i$$
$$M_a = 113.5102$$

Step 5. Compute the density of the oil system:

$$\rho_0 = \frac{M_a}{V_m}$$

$$\rho_0 = \frac{113.5102}{2.528417} = 44.896 \text{ lb/ft}^3$$

Equations of State

An equation of state (EOS) is an analytical expression relating the pressure p to the temperature T and the volume V. A proper description of this PVT relationship for real hydrocarbon fluids is essential in determining the volumetric and phase behavior of petroleum reservoir fluids and in predicting the performance of surface separation facilities.

The best known and the simplest example of an equation of state is the ideal gas equation, expressed mathematically by the expression

$$p = \frac{RT}{V} \tag{7-1}$$

where V = gas volume in cubic feet per 1 mol of gas. This PVT relationship is used to describe the volumetric behavior of real hydrocarbon gases only at pressures close to the atmospheric pressure for which it was experimentally derived.

The extreme limitations of the applicability of Equation 7-1 prompted numerous attempts to develop an equation of state suitable for describing the behavior of real fluids at extended ranges of pressures and temperatures.

The main objective of this chapter is to review developments and advances in the field of empirical cubic equations of state and demonstrate their applications in petroleum engineering.

7.1 THE VAN DER WAALS EQUATION OF STATE

In developing the ideal gas EOS (Equation 7-1), two assumptions were made:

- **First assumption**. The volume of the gas molecules is insignificant compared to the volume of the container and the distance between the molecules.

- **Second assumption.** There are no attractive or repulsive forces between the molecules or the walls of the container.

van der Waals (1873) [61] attempted to eliminate these two assumptions by developing an empirical equation of state for real gases. In his attempt to eliminate the first assumption, van der Waals pointed out that the gas molecules occupy a significant fraction of the volume at higher pressures and proposed that the volume of the molecules, as denoted by the parameter b, be subtracted from the actual molar volume V in Equation 7-1, to give

$$p = \frac{RT}{V - b}$$

where the parameter b is known as the covolume and is considered to reflect the volume of molecules. The variable V represents the actual volume in cubic feet per 1 mol of gas.

To eliminate the second assumption, van der Waals subtracted a corrective term, denoted by a/V^2, from the preceding equation to account for the attractive forces between molecules. In a mathematical form, van der Waals proposed the following expression:

$$p = \frac{RT}{V - b} - \frac{a}{V^2} \tag{7-2}$$

where p = system pressure, psia
     T = system temperature, °R
     R = gas constant, 10.73 psi-ft^3/lb-mol = °R
     V = volume, ft^3/mol

The two parameters a and b are constants characterizing the molecular properties of the individual components. The symbol a is considered a measure of the intermolecular attractive forces between the molecules. Equation 7-2 shows the following important characteristics:

1. At low pressures, the volume of the gas phase is large in comparison with the volume of the molecules. The parameter b becomes negligible in comparison with V and the attractive forces term a/V^2 becomes insignificant; therefore, the van der Waals equation reduces to the ideal gas equation (Equation 7-1).
2. At high pressure, i.e., $p \rightarrow \infty$, volume V becomes very small and approaches the value b, which is the actual molecular volume.

The van der Waals or any other equation of state can be expressed in a more generalized form as follows:

$$p = p_{\text{repulsive}} - p_{\text{attractive}}$$

where the repulsive pressure term $p_{repulsive}$ is represented by the term $RT/(V-b)$ and the attractive pressure term $p_{attractive}$ is described by a/V_2.

In determining the values of the two constants a and b for any pure substance, van der Waals observed that the critical isotherm has a horizontal slope and an inflection point at the critical point, as shown in Figure 7-1. This observation can be expressed mathematically as follows:

$$\left[\frac{\partial p}{\partial V}\right]_{T_C, p_C} = 0, \qquad \left[\frac{\partial^2 p}{\partial V^2}\right]_{T_C, p_C} = 0 \qquad (7\text{-}3)$$

Differentiating Equation 7-2 with respect to the volume at the critical point results in

$$\left[\frac{\partial p}{\partial V}\right]_{T_C, p_C} = \frac{-RT_C}{(V_C - b)^3} + \frac{2a}{V_C^3} = 0 \qquad (7\text{-}4)$$

$$\left[\frac{\partial^2 p}{\partial V^2}\right]_{T_C, p_C} = \frac{2RT_C}{(V_C - b)^3} + \frac{6a}{V_C^4} = 0 \qquad (7\text{-}5)$$

Solving Equations 7-4 and 7-5 simultaneously for the parameters a and b gives

$$b = \left(\frac{1}{3}\right)V_C \qquad (7\text{-}6)$$

$$a = \left(\frac{8}{9}\right)RT_C V_C \qquad (7\text{-}7)$$

Equation 7-6 suggests that the volume of the molecules b is approximately 0.333 of the critical volume V_C of the substance. Experimental studies reveal that the covolume b is in the range of 0.24 to 0.28 of the critical volume and pure component.

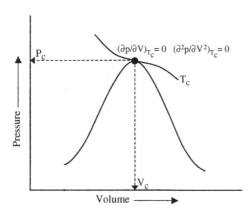

FIGURE 7-1 An idealized pressure–volume relationship for a pure compound.

By applying Equation 7-2 to the critical point (i.e., by setting $T = T_c$, $p = p_c$, and $V = V_c$) and combining with Equations 7-6 and 7-7, we get

$$p_C V_C = (0.375)RT_C \qquad (7\text{-}8)$$

Equation 7-8 shows that regardless of the type of substance, the van der Waals EOS produces a universal critical gas compressibility factor Z_c of 0.375. Experimental studies show that Z_c values for substances range between 0.23 and 0.31.

Equation 7-8 can be combined with Equations 7-6 and 7-7 to give a more convenient and traditional expression for calculating the parameters a and b to yield

$$a = \Omega_a \frac{R^2 T_c^2}{p_c} \qquad (7\text{-}9)$$

$$b = \Omega_b \frac{RT_c}{p_c} \qquad (7\text{-}10)$$

where R = gas constant, 10.73 psia-ft^3/lb-mol-°R
  p_c = critical pressure, psia
  T_c = critical temperature, °R
  Ω_a = 0.421875
  Ω_b = 0.125

Equation 7-2 can also be expressed in a cubic form in terms of the volume V as follows:

$$V^3 - \left(b + \frac{RT}{p}\right)V^2 + \left(\frac{a}{p}\right)V - \left(\frac{ab}{p}\right) = 0 \qquad (7\text{-}11)$$

Equation 7-11 is usually referred to as the *van der Waals two-parameter cubic equation of state*. The term *two-parameter* refers to the parameters a and b. The term *cubic equation of state* implies an equation that, if expanded, would contain volume terms to the first, second, and third power.

Perhaps the most significant feature of Equation 7-11 is its ability to describe the liquid-condensation phenomenon and the passage from the gas to the liquid phase as the gas is compressed. This important feature of the van der Waals EOS is discussed in conjunction with Figure 7-2.

Consider a pure substance with a p-V behavior as shown in Figure 7-2. Assume that the substance is kept at a constant temperature T below its critical temperature. At this temperature, Equation 7-11 has three real roots (volumes) for each specified pressure p. A typical solution of Equation 7-11 at constant temperature T is shown graphically by the dashed

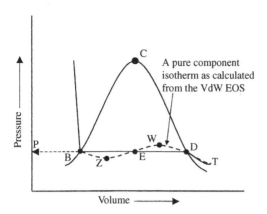

FIGURE 7-2 Pressure–volume diagram for a pure component.

isotherm: the constant temperature curve DWEZB in Figure 7-2. The three values of V are the intersections B, E, and D on the horizontal line, corresponding to a fixed value of the pressure. This dashed calculated line (DWEZB) then appears to give a continuous transition from the gaseous phase to the liquid phase, but in reality the transition is abrupt and discontinuous, with both liquid and vapor existing along the straight horizontal line DB. Examining the graphical solution of Equation 7-11 shows that the largest root (volume), as indicated by point D, corresponds to the volume of the saturated vapor, while the smallest positive volume, as indicated by point B, corresponds to the volume of the saturated liquid. The third root, point E, has no physical meaning. Note that these values become identical as the temperature approaches the critical temperature T_c of the substance.

Equation 7-11 can be expressed in a more practical form in terms of the compressibility factor Z. Replacing the molar volume V in Equation 7-11 with ZRT/p gives

$$Z^3 - (1 + B)Z^2 + AZ - AB = 0 \qquad (7\text{-}12)$$

where

$$A = \frac{ap}{R^2T^2} \qquad (7\text{-}13)$$

$$B = \frac{bp}{RT} \qquad (7\text{-}14)$$

Z = compressibility factor
p = system pressure, psia
T = system temperature, °R

Equation 7-12 yields one real root[1] in the one-phase region and three real roots in the two-phase region (where the system pressure equals the vapor pressure of the substance). In the latter case, the largest root corresponds to the compressibility factor of the vapor phase Z^V, while the smallest positive root corresponds to that of the liquid Z^L.

An important practical application of Equation 7-12 is calculating density, as illustrated in the following example.

Example 7-1

A pure propane is held in a closed container at 100°F. Both gas and liquid are present. Calculate, using the van der Waals EOS, the density of the gas and liquid phases.

Solution

Step 1. Determine the vapor pressure p_v of the propane from the Cox chart (Figure 1-1). This is the only pressure at which two phases can exist at the specified temperature:

$$p_v = 185 \text{ psi}$$

Step 2. Calculate parameters a and b from Equations 7-9 and 7-10, respectively:

$$a = \Omega_a \frac{R^2 T_c^2}{P_c}$$

$$a = 0.421875 \frac{(10.73)^2 (666)^2}{616.3} = 34,957.4$$

and

$$b = \Omega_b \frac{R T_c}{P_c}$$

$$b = 0.125 \frac{10.73(666)}{616.3} = 1.4494$$

---

[1]In some supercritical regions, Equation 7-12 can yield three real roots for Z. From the three real roots, the largest root is the value of the compressibility with physical meaning.

Step 3. Compute coefficients A and B by applying Equations 7-13 and 7-14, respectively.

$$A = \frac{ap}{R^2 T^2}$$

$$A = \frac{(34,957.4)(185)}{(10.73)^2 (560)^2} = 0.179122$$

$$B = \frac{bp}{RT}$$

$$B = \frac{(1.4494)(185)}{(10.73)(560)} = 0.044625$$

Step 4. Substitute the values of A and B into Equation 7-12 to give

$$Z^3 - (1+B)Z^2 + AZ - AB = 0$$
$$Z^3 - 1.044625Z^2 + 0.179122Z - 0.007993 = 0$$

Step 5. Solve the preceding third-degree polynomial by extracting the largest and smallest roots of the polynomial using the appropriate direct or iterative method to give

$$Z^v = 0.72365$$
$$Z^L = 0.07534$$

Step 6. Solve for the density of the gas and liquid phases:

$$\rho_g = \frac{pM}{Z^v RT}$$

$$\rho_g = \frac{(185)(44.0)}{(0.72365)(10.73)(560)} = 1.87 \text{ lb/ft}^3$$

and

$$\rho_L = \frac{pM}{Z^L RT}$$

$$\rho_L = \frac{(185)(44)}{(0.7534)(10.73)(560)} = 17.98 \text{ lb/ft}^3$$

The van der Waals equation of state, despite its simplicity, provides a correct description, at least qualitatively, of the PVT behavior of substances in the liquid and gaseous states. Yet it is not accurate enough to be suitable for design purposes.

With the rapid development of computers, the EOS approach for the calculation of physical properties and phase equilibria proved to be a powerful tool, and much energy was devoted to the development of new and accurate equations of state. These equations, many of them a modification of the van der Waals equation of state, range in complexity from simple expressions containing 2 or 3 parameters to complicated forms containing more than 50 parameters. Although the complexity of any equation of state presents no computational problem, most authors prefer to retain the simplicity found in the van der Waals cubic equation while improving its accuracy through modifications.

All equations of state are generally developed for pure fluids first, then extended to mixtures through the use of mixing rules. These mixing rules are simply means of calculating mixture parameters equivalent to those of pure substances.

7.2 REDLICH-KWONG EQUATION OF STATE

Redlich and Kwong (1949) [45] demonstrated that, by a simple adjustment, the van der Waals attractive pressure term a/V^2 could considerably improve the prediction of the volumetric and physical properties of the vapor phase. The authors replaced the attractive pressure term with a generalized temperature dependence term. Their equation has the following form:

$$p = \frac{RT}{V-b} - \frac{a}{V(V+b)\sqrt{T}} \tag{7-15}$$

where T is the system temperature in °R.

Redlich and Kwong, in their development of the equation, noted that, as the system pressure becomes very large, i.e., $p \rightarrow \infty$, the molar volume V of the substance shrinks to about 26% of its critical volume regardless of the system temperature. Accordingly, they constructed Equation 7-15 to satisfy the following condition:

$$b = 0.26\,V_c \tag{7-16}$$

Imposing the critical point conditions (as expressed by Equation 7-3) on Equation 7-15 and solving the resulting equations simultaneously gives

$$a = \Omega_a \frac{R^2 T_c^{2.5}}{P_c} \tag{7-17}$$

$$b = \Omega_b \frac{RT_c}{P_c} \tag{7-18}$$

where $\Omega_a = 0.42747$ and $\Omega_b = 0.08664$. Equating Equation 7-18 with 7-16 gives

$$p_c V_c = 0.333 R T_c \qquad (7\text{-}19)$$

Equation 7-19 shows that the Redlich-Kwong EOS produces a universal critical compressibility factor (Z_c) of 0.333 for all substances. As indicated earlier, the critical gas compressibility ranges from 0.23 to 0.31 for most of the substances.

Replacing the molar volume V in Equation 7-15 with ZRT/p gives

$$Z^3 - Z^2 + (A - B - B^2)Z - AB = 0 \qquad (7\text{-}20)$$

where

$$A = \frac{ap}{R^2 T^{2.5}} \qquad (7\text{-}21)$$

$$B = \frac{bp}{RT} \qquad (7\text{-}22)$$

As in the van der Waals EOS, Equation 7-20 yields one real root in the one-phase region (gas-phase region or liquid-phase region) and three real roots in the two-phase region. In the latter case, the *largest root* corresponds to the compressibility factor of the gas phase Z^v while the *smallest positive root* corresponds to that of the liquid Z^L.

Example 7-2

Rework Example 7-1 by using the Redlich-Kwong equation of state.

Solution

Step 1. Calculate the parameters a, b, A, and B:

$$a = 0.42747 \frac{(10.73)^2 (666)^{2.5}}{616.3} = 914{,}110.1$$

$$b = 0.08664 \frac{(10.73)(666)}{616.3} = 1.0046$$

$$A = \frac{(914{,}110.1)(185)}{(10.73)^2 (560)^{2.5}} = 0.197925$$

$$B = \frac{(1.0046)(185)}{(10.73)(560)} = 0.03093$$

Step 2. Substitute parameters A and B into Equation 7-20, and extract the largest and smallest root, to give

$$Z^3 - Z^2 + 0.1660384Z - 0.0061218 = 0$$
Largest root $Z^v = 0.802641$
Smallest root $Z^L = 0.0527377$

Step 3. Solve for the density of the liquid phase and the gas phase:

$$\rho = \frac{pM}{ZRT}$$

$$\rho^L = \frac{(185)(44)}{(0.0527377)(10.73)(560)} = 25.7 \text{ lb/ft}^3$$

$$\rho^v = \frac{(185)(44)}{(0.802641)(10.73)(560)} = 1.688 \text{ lb/ft}^3$$

Redlich and Kwong extended the application of their equation to hydrocarbon liquid or gas mixtures by employing the following mixing rules:

$$a_m = \left[\sum_{i=1}^{n} x_i\sqrt{a_i}\right]^2 \qquad (7\text{-}23)$$

$$b_m = \sum_{i=1}^{n}[x_i b_i] \qquad (7\text{-}24)$$

where n = number of components in mixture
a_i = Redlich-Kwong a parameter for the ith component as given by Equation 7-17
b_i = Redlich-Kwong b parameter for the ith component as given by Equation 7-18
a_m = parameter a for mixture
b_m = parameter b for mixture
x_i = mole fraction of component i in the liquid phase

To calculate a_m and b_m for a hydrocarbon gas mixture with a composition of y_i, use Equations 7-23 and 7-24 and replace x_i with y_i:

$$a_m = \left[\sum_{i=1}^{n} y_i\sqrt{a_i}\right]^2$$

$$b_m = \sum_{i=1}^{n}[y_i b_i]$$

Equation 7-20 gives the compressibility factor of the gas phase or the liquid phase with the coefficients A and B as defined by Equations 7-21 and 7-22. The application of the Redlich-Kwong equation of state for hydrocarbon mixtures can be best illustrated through the following two examples.

Example 7-3

Calculate the density of a crude oil with the following composition at 4000 psia and 160°F. Use the Redlich-Kwong EOS.

Component	x_i	M	p_c	T_c
C_1	0.45	16.043	666.4	343.33
C_2	0.05	30.070	706.5	549.92
C_3	0.05	44.097	616.0	666.06
$n - C_4$	0.03	58.123	527.9	765.62
$n - C_5$	0.01	72.150	488.6	845.8
C_6	0.01	84.00	453	923
C_{7+}	0.40	215	285	1287

Solution

Step 1. Determine the parameters a_i and b_i for each component using Equations 7-17 and 7-18:

Component	a_i	b_i
C_1	161,044.3	0.4780514
C_2	493,582.7	0.7225732
C_3	914,314.8	1.004725
C_4	1,449,929	1.292629
C_5	2,095,431	1.609242
C_6	2,845,191	1.945712
C_{7+}	1.022348E7	4.191958

Step 2. Calculate the mixture parameters a_m and b_m from Equations 7-23 and 7-24 to give

$$a_m = \left[\sum_{i=1}^{n} x_i \sqrt{a_i} \right]^2 = 2,591,967$$

and

$$b_m = \sum_{i=1}^{n} [x_i b_i] = 2.0526$$

Step 3. Compute the coefficients A and B using Equations 7-21 and 7-22 to produce

$$A = \frac{a_m p}{R^2 T^{2.5}} = \frac{2,591,967(4000)}{10.73^2(620)^{2.5}} = 9.406539$$

$$B = \frac{b_m p}{RT} = \frac{2.0526(4000)}{10.73(620)} = 1.234049$$

Step 4. Solve Equation 7-20 for the largest positive root to yield

$$Z^3 - Z^2 + 6.93845Z - 11.60813 = 0$$
$$Z^L = 1.548126$$

Step 5. Calculate the apparent molecular weight of the crude oil:

$$M_a = \sum x_i M_i$$
$$M_a = 100.2547$$

Step 6. Solve for the density of the crude oil:

$$\rho^L = \frac{p M_a}{Z^L RT}$$

$$\rho^L = \frac{(4000)(100.2547)}{(10.73)(620)(1.548120)} = 38.93 \ lb/ft^3$$

Notice that liquid density, as calculated by Standing's correlation, gives a value of 46.23 lb/ft³.

Example 7-4

Calculate the density of a gas phase with the following composition at 4000 psia and 160°F. Use the Redlich-Kwong EOS.

Component	y_i	M	p_c	T_c
C_1	0.86	16.043	666.4	343.33
C_2	0.05	30.070	706.5	549.92
C_3	0.05	44.097	616.0	666.06
C_4	0.02	58.123	527.9	765.62
C_5	0.01	72.150	488.6	845.8
C_6	0.005	84.00	453	923
C_{7+}	0.005	215	285	1287

Solution

Step 1. Calculate a_m and b_m using Equations 7-23 and 7-24 to give

$$a_m = \left[\sum_{i=1}^{n} y_i \sqrt{a_i}\right]^2$$

$$a_m = 241{,}118$$

$$b_m = \sum b_i x_i$$

$$b_m = 0.5701225$$

Step 2. Calculate the coefficients A and B by applying Equations 7-21 and 7-22 to yield

$$A = \frac{a_m p}{R^2 T^{2.5}} = \frac{241{,}118(4000)}{10.73^2 (620)^{2.5}} = 0.8750$$

$$B = \frac{b_m p}{RT} = \frac{0.5701225(4000)}{10.73(620)} = 0.3428$$

Step 3. Solve Equation 7-20 for Z^V to give

$$Z^3 - Z^2 + 0.414688Z - 0.29995 = 0$$

$$Z^V = 0.907$$

Step 4. Calculate the apparent density of the gas mixture:

$$M_a = \sum y_i M_i = 20.89$$

$$\rho^v = \frac{p M_a}{Z^v RT}$$

$$\rho^v = \frac{(4000)(20.89)}{(10.73)(620)(0.907)} = 13.85 \text{ lb/ft}^3$$

7.3 SOAVE-REDLICH-KWONG EQUATION OF STATE AND ITS MODIFICATIONS

One of the most significant milestones in the development of cubic equations of state was the publication by Soave (1972) [54] of a modification to the evaluation of parameter a in the attractive pressure term of the Redlich-Kwong equation of state (Equation 7-15). Soave replaced the term $a/T^{0.5}$ in Equation 7-15 with a more generalized *temperature-dependent term*, as denoted by (aα), to give

$$p = \frac{RT}{V-b} - \frac{a\alpha}{V(V+b)} \tag{7-25}$$

where α is a dimensionless factor that becomes unity at $T = T_c$. At temperatures other than critical temperature, the parameter α is defined by the following expression:

$$\alpha = \left[1 + m\left(1 - \sqrt{T_r}\right)\right]^2 \tag{7-26}$$

The parameter m is correlated with the acentric factor to give

$$m = 0.480 + 1.574\omega - 0.176\omega^2 \tag{7-27}$$

where T_r = reduced temperature T/T_c
    ω = acentric factor of the substance
    T = system temperature, °R

For any pure component, the constants a and b in Equation 7-25 are found by imposing the classical van der Waals critical point constraints (Equation 7-3) on Equation 7-25 and solving the resulting equations to give

$$a = \Omega_a \frac{R^2 T_c^2}{P_c} \tag{7-28}$$

$$b = \Omega_b \frac{RT_c}{P_c} \tag{7-29}$$

where Ω_a and Ω_b are the Soave-Redlich-Kwong (SRK) dimensionless pure component parameters and have the following values:

$$\Omega_a = 0.42747 \quad \text{and} \quad \Omega_b = 0.08664$$

Edmister and Lee (1986) [13] showed that the two parameters a and b can be determined more conveniently by considering the critical isotherm:

$$(V - V_c)^3 = V^3 - [3V_c]V^2 + [3V_c^2]V - V_c^3 = 0 \tag{7-30}$$

Equation 4-11 can also be put into a cubic form to give

$$V^3 - \left[\frac{RT}{P}\right]V^2 + \left[\frac{a\alpha}{P} - \frac{bRT}{P} - b^2\right]V - \left[\frac{(a\alpha)b}{P}\right] = 0 \tag{7-31}$$

At the critical point, the coefficient $\alpha = 1$ and the preceding two expressions are essentially identical. Equating the like terms gives

$$3V_c = \frac{RT_c}{P_c} \tag{7-32}$$

$$3V_c^2 = \frac{a}{P_c} - \frac{bRT_c}{P_c} - b^2 \qquad (7\text{-}33)$$

and

$$V_c^3 = \frac{ab}{P_c} \qquad (7\text{-}34)$$

Solving these equations for parameters a and b yields expressions for the parameters as given by Equations 7-28 and 7-29.

Equation 7-32 indicates that the SRK equation of state gives a universal critical gas compressibility factor of 0.333. Combining Equation 6-3 with 7-29 gives

$$b = 0.26V_c$$

Introducing the compressibility factor Z into Equation 6-2 by replacing the molar volume V in the equation with (ZRT/p) and rearranging gives

$$Z^3 - Z^2 + (A - B - B^2)Z - AB = 0 \qquad (7\text{-}35)$$

with

$$A = \frac{(a\alpha)p}{(RT)^2} \qquad (7\text{-}36)$$

$$B = \frac{bp}{RT} \qquad (7\text{-}37)$$

where p = system pressure, psia
   T = system temperature, °R
   R = 10.730 psia ft^3/lb-mol-°R

Example 7-5

Rework Example 7-1 and solve for the density of the two phases by using the SRK EOS.

Solution

Step 1. Determine the critical pressure, critical temperature, and acentric factor to give

$$T_c = 666.01°R$$
$$P_c = 616.3 \text{ psia}$$
$$\omega = 0.1524$$

Step 2. Calculate the reduced temperature:

$$T_r = 560/666.01 = 0.8408$$

Step 3. Calculate the parameter m by applying Equation 7-27 to yield

$$m = 0.480 + 1.574\omega - 0.176\omega^2$$
$$m = 0.480 + 1.574(0.1524) - 0.176(1.524)^2 = 0.7051$$

Step 4. Solve for the parameter a by using Equation 7-26 to give

$$\alpha = \left[m + \left(1 - \sqrt{T_r} \right) \right]^2 = 1.120518$$

Step 5. Compute the coefficients a and b by applying Equations 7-28 and 7-29 to yield

$$a = 0.42747 \frac{10.73^2(666.01)^2}{616.3} = 35,427.6$$

$$b = 0.08664 \frac{10.73(666.01)}{616.3} = 1.00471$$

Step 6. Calculate the coefficients A and B from Equations 7-36 and 7-37 to produce

$$A = \frac{(a\alpha)p}{R^2T^2}$$

$$A = \frac{(35,427.6)(1.120518)185}{10.73^2(560)^2} = 0.203365$$

$$B = \frac{bp}{RT}$$

$$B = \frac{(1.00471)(185)}{(10.73)(560)} = 0.034658$$

Step 7. Solve Equation 7-35 for Z^L and Z^v:

$$Z^3 - Z^2 + (A - B - B^2)Z + AB = 0$$
$$Z^3 - Z^2 + (0.203365 - 0.034658 - 0.034658^2)Z$$
$$+ (0.203365)(0.034658) = 0$$

Solving the above third-degree polynomial gives

$$Z^L = 0.06729$$
$$Z^V = 0.80212$$

Step 8. Calculate the gas and liquid density to give

$$\rho = \frac{pM}{ZRT}$$

$$\rho^v = \frac{(185)(44.0)}{(0.802121)(10.73)(560)} = 1.6887 \text{ lb/ft}^3$$

$$\rho^L = \frac{(185)(44.0)}{(0.06729)(10.73)(560)} = 20.13 \text{ lb/ft}^3$$

To use Equation 7-35 with mixtures, mixing rules are required to determine the terms (aα) and b for the mixtures. Soave adopted the following mixing rules:

$$(a\alpha)_m = \sum_i \sum_j \left[x_i x_j \sqrt{a_i a_j \alpha_i \alpha_j} (1 - k_{ij}) \right] \qquad (7\text{-}38)$$

$$b_m = \sum_i [x_i b_i] \qquad (7\text{-}39)$$

with

$$A = \frac{(a\alpha)_m p}{(RT)^2} \qquad (7\text{-}40)$$

and

$$B = \frac{b_m p}{RT} \qquad (7\text{-}41)$$

The parameter k_{ij} is an empirically determined correction factor (called the *binary interaction coefficient*) that is designed to characterize any binary system formed by component i and component j in the hydrocarbon mixture.

These binary interaction coefficients are used to model the intermolecular interaction through empirical adjustment of the $(a\alpha)_m$ term as represented mathematically by Equation 7-38. They are dependent on the difference in molecular size of components in a binary system, and they are characterized by the following properties:

- The interaction between hydrocarbon components increases as the relative difference between their molecular weights increases:

$$k_{i,j+1} > k_{i,j}$$

- Hydrocarbon components with the same molecular weight have a binary interaction coefficient of 0:

$$k_{i,i} = 0$$

- The binary interaction coefficient matrix is symmetric:

$$k_{j,i} = k_{i,j}$$

Slot-Peterson (1987) [53] and Vidal and Daubert (1978) [62] presented a theoretical background to the meaning of the interaction coefficient and techniques for determining their values. Grabowski and Daubert (1978) [17] and Soave (1972) [54] suggested that no binary interaction coefficients are required for hydrocarbon systems. However, with nonhydrocarbons present, binary interaction parameters can greatly improve the volumetric and phase behavior predictions of the mixture by the SRK EOS.

In solving Equation 7-30 for the compressibility factor of the liquid phase, the composition of the liquid x_i is used to calculate the coefficients A and B of Equations 7-40 and 7-41 through the use of the mixing rules as described by Equations 7-38 and 7-39. For determining the compressibility factor of the gas phase Z^v, the previously outlined procedure is used with composition of the gas phase y_i replacing x_i.

Example 7-6

A two-phase hydrocarbon system exists in equilibrium at 4000 psia and 160°F. The system has the following composition:

Component	x_i	y_i
C_1	0.45	0.86
C_2	0.05	0.05
C_3	0.05	0.05
C_4	0.03	0.02
C_5	0.01	0.01
C_6	0.01	0.005
C_{7+}	0.40	0.005

The heptanes-plus fraction has the following properties:

$$M = 215$$
$$P_c = 285 \text{ psia}$$
$$T_c = 700°F$$
$$\omega = 0.52$$

Assuming $k_{ij} = 0$, calculate the density of each phase using the SRK EOS.

Solution

Step 1. Calculate the parameters α, a, and b by applying Equations 7-21, 7-28, and 7-29:

Component	α_i	a_i	b_i
C_1	0.6869	8,689.3	0.4780
C_2	0.9248	21,040.8	0.7725
C_3	1.0502	35,422.1	1.0046
C_4	1.1616	52,390.3	1.2925
C_5	1.2639	72,041.7	1.6091
C_6	1.3547	94,108.4	1.9455
C_{7+}	1.7859	232,367.9	3.7838

Step 2. Calculate the mixture parameters $(a\alpha)_m$ and b_m for the gas phase and liquid phase by applying Equations 7-38 and 7-39 to give

- **For the gas phase using y_i.**

$$(a\alpha)_m = \sum_i \sum_j \left[y_i y_j \sqrt{a_i a_j \alpha_i \alpha_j} \left(1 - k_{ij}\right) \right] = 9219.3$$

$$b_m = \sum_i [y_i b_i] = 0.5680$$

- **For the liquid phase using x_i.**

$$(a\alpha)_m = \sum_i \sum_j \left[x_i x_j \sqrt{a_i a_j \alpha_i \alpha_j} \left(1 - k_{ij}\right) \right] = 104{,}362.9$$

$$b_m = \sum_i [x_i b_i] = 0.1.8893$$

Step 3. Calculate the coefficients A and B for each phase by applying Equations 7-40 and 7-41 to yield

- **For the gas phase.**

$$A = \frac{(a\alpha)_m P}{R^2 T^2} = \frac{(9219.3)(4000)}{(10.73)^2 (620)^2} = 0.8332$$

$$B = \frac{b_m P}{RT} = \frac{(0.5680)(4000)}{(10.73)(620)} = 0.3415$$

- **For the liquid phase.**

$$A = \frac{(a\alpha)_m P}{R^2 T^2} = \frac{(104{,}362.9)(4000)}{(10.73)^2 (620)^2} = 9.4324$$

$$B = \frac{b_m P}{RT} = \frac{(1.8893)(4000)}{(10.73)(620)} = 1.136$$

Step 4. Solve Equation 7-35 for the compressibility factor of the gas phase to produce

$$Z^3 - Z^2 + (A - B - B^2)Z + AB = 0$$
$$Z^3 - Z^2 + (0.8332 - 0.3415 - 0.3415^2)Z + (0.8332)(0.3415) = 0$$

Solving this polynomial for the largest root gives

$$Z^v = 0.9267$$

Step 5. Solve Equation 7-35 for the compressibility factor of the liquid phase to produce

$$Z^3 - Z^2 + (A - B - B^2)Z + AB = 0$$
$$Z^3 - Z^2 + (9.4324 - 1.136 - 1.136^2)Z + (9.4324)(1.136) = 0$$

Solving this polynomial for the smallest root gives

$$Z^L = 1.4121$$

Step 6. Calculate the apparent molecular weight of the gas phase and liquid phase from their composition to yield

- **For the gas phase.**

$$M_a = \sum y_i M_i = 20.89$$

- **For the liquid phase.**

$$M_a = \sum x_i M_i = 100.25$$

Step 7. Calculate the density of each phase:

$$\rho = \frac{pM_a}{RTZ}$$

- **For the gas phase.**

$$\rho^v = \frac{(4000)(20.89)}{(10.73)(620)(0.9267)} = 13.556 \text{ lb/ft}^3$$

- **For the liquid phase.**

$$\rho^L = \frac{(4000)(100.25)}{(10.73)(620)(1.4121)} = 42.68 \text{ lb/ft}^3$$

It is appropriate at this time to introduce and define the concept of the fugacity and the fugacity coefficient of the component. The *fugacity* f is a measure of the molar Gibbs energy of a real gas. It is evident from the definition that the fugacity has the units of pressure; in fact, the fugacity may be looked on as a vapor pressure modified to correctly represent the escaping tendency of the molecules from one phase into the other. In mathematical form, the fugacity of a pure component is defined by the following expression:

$$f = p \exp\left[\int_0^P \left(\frac{Z-1}{p}\right) dp\right] \tag{7-42}$$

where f = fugacity, psia
$\quad p$ = pressure, psia
$\quad Z$ = compressibility factor

The ratio of the fugacity to the pressure, i.e., f/p, is called the *fugacity coefficient* Φ and is calculated from Equation 7-42 as

$$\frac{f}{p} = \Phi = \exp\left[\int_0^P \left(\frac{Z-1}{p}\right) dp\right]$$

Soave applied this generalized thermodynamic relationship to Equation 7-25 to determine the fugacity coefficient of a pure component:

$$\ln\left(\frac{f}{p}\right) = \ln(\Phi) = Z - 1 - \ln(Z - B) - \frac{A}{B}\ln\left[\frac{Z+B}{Z}\right] \tag{7-43}$$

In practical petroleum engineering applications, we are concerned with the phase behavior of the hydrocarbon liquid mixture, which, at a specified pressure and temperature, is in equilibrium with a hydrocarbon gas mixture at the same pressure and temperature.

The component fugacity in each phase is introduced to develop a criterion for thermodynamic equilibrium. Physically, the fugacity of a component i in one phase with respect to the fugacity of the component in a second phase is a measure of the potential for transfer of the component between phases. The phase with a lower component fugacity accepts the component from the phase with a higher component fugacity. Equal fugacities of a component in the two phases result in a zero net transfer. A zero transfer for all components implies a hydrocarbon system that is in thermodynamic equilibrium. Therefore, the condition of the thermodynamic equilibrium can be expressed mathematically by

$$f_i^v = f_i^L \quad 1 \le i \le n \tag{7-44}$$

where f_i^v = fugacity of component i in the gas phase, psi
$\quad f_i^L$ = fugacity of component i in the liquid phase, psi
$\quad n$ = number of components in the system

The fugacity coefficient of component i in a hydrocarbon liquid mixture or hydrocarbon gas mixture is a function of

- System pressure.
- Mole fraction of the component.
- Fugacity of the component.

For a component i in the gas phase, the fugacity coefficient is defined as

$$\Phi_i^v = \frac{f_i^v}{y_i p} \tag{7-45}$$

For a component i in the liquid phase, the fugacity coefficient is

$$\Phi_i^L = \frac{f_i^L}{x_i p} \tag{7-46}$$

where Φ_i^v = fugacity coefficient of component i in the vapor phase

Φ_i^L = fugacity coefficient of component i in the liquid phase

It is clear that, at equilibrium $f_i^L = f_i^v$, the equilibrium ratio K_i as previously defined by Equation 1-1, i.e., $K_i = y_i/x_i$, can be redefined in terms of the fugacity of components as

$$K_i = \frac{\left[f_i^L/(x_i p) \right]}{\left[f_i^v/(y_i p) \right]} = \frac{\Phi_i^L}{\Phi_i^v} \tag{7-47}$$

Reid et al. (1977) [46] defined the fugacity coefficient of component i in a hydrocarbon mixture by the following generalized thermodynamic relationship:

$$\ln(\Phi_i) = \left(\frac{1}{RT} \right) \left[\int_v^\infty \left(\frac{\partial p}{\partial n_i} - \frac{RT}{V} \right) dV \right] - \ln(Z) \tag{7-48}$$

where V = total volume of n models of the mixture

n_i = number of moles of component i

Z = compressibility factor of the hydrocarbon mixture

By combining this thermodynamic definition of the fugacity with the SRK EOS (Equation 7-25), Soave proposed the following expression for the fugacity coefficient of component i in the liquid phase:

$$\ln(\Phi_i^L) = \frac{b_i(Z^L - 1)}{b_m} - \ln(Z^L - B) - \left(\frac{A}{B} \right) \left[\frac{2\Psi_i}{(a\alpha)_m} - \frac{b_i}{b_m} \right] \ln \left[1 + \frac{B}{Z^L} \right] \tag{7-49}$$

where

$$\Psi_i = \sum_j \left[x_j \sqrt{a_i a_j \alpha_i \alpha_j} (1 - k_{ij}) \right] \tag{7-50}$$

$$(a\alpha)_m = \sum_i \sum_j \left[x_i x_j \sqrt{a_i a_j \alpha_i \alpha_j} (1 - k_{ij}) \right] \qquad (7\text{-}51)$$

Equation 7-49 is also used to determine the fugacity coefficient of component i in the gas phase Φ_i^v by using the composition of the gas phase y_i in calculating A, B, Z^v, and other composition-dependent terms, or

$$\ln(\Phi_i^v) = \frac{b_i(Z^v - 1)}{b_m} - \ln(Z^v - B) - \left(\frac{A}{B}\right)\left[\frac{2\Psi_i}{(a\alpha)_m} - \frac{b_i}{b_m}\right] \ln\left[1 + \frac{B}{Z^v}\right]$$

where

$$\Psi_i = \sum_j \left[y_j \sqrt{a_i a_j \alpha_i \alpha_j} (1 - k_{ij}) \right]$$
$$(a\alpha)_m = \sum_i \sum_j \left[y_i y_j \sqrt{a_i a_j \alpha_i \alpha_j} (1 - k_{ij}) \right]$$

7.4 MODIFICATIONS OF THE SRK EQUATION OF STATE

To improve the pure component vapor pressure predictions by the SRK equation of state, Grabowski and Daubert (1978) [17] proposed a new expression for calculating parameter m of Equation 7-27. The proposed relationship originated from analyses of extensive experimental data for pure hydrocarbons. The relationship has the following form:

$$m = 0.48508 + 1.55171\omega - 0.15613\omega^2 \qquad (7\text{-}52)$$

Sim and Daubert (1980) [52] pointed out that, because the coefficients of Equation 7-52 were determined by analyzing vapor pressure data of low-molecular-weight hydrocarbons, it is unlikely that Equation 7-52 will suffice for high-molecular-weight petroleum fractions. Realizing that the acentric factors for the heavy petroleum fractions are calculated from an equation such as the Edmister correlation or the Lee and Kessler (1975) correlation, the authors proposed the following expressions for determining the parameter m:

- If the acentric factor is determined using the Edmister correlation, then

$$m = 0.431 + 1.57\omega_i - 0.161\omega_i^2 \qquad (7\text{-}53)$$

- If the acentric factor is determined using the Lee and Kessler correction, then

$$m = 0.315 + 1.60\omega_i - 0.166\omega_i^2 \qquad (7\text{-}54)$$

Elliot and Daubert (1985) [14] stated that the optimal binary interaction coefficient k_{ij} would minimize the error in the representation of all thermodynamic properties of a mixture. Properties of particular interest in phase equilibrium calculations include bubble-point pressure, dew-point pressure, and equilibrium ratios. The authors proposed a set of relationships for determining interaction coefficients for asymmetric mixtures[2] that contain methane, N_2, CO_2, and H_2S. Referring to the principal component as i and the other fraction as j, Elliot and Daubert proposed the following expressions:

- **For N_2 systems.**

$$k_{ij} = 0.107089 + 2.9776k_{ij}^{\infty} \qquad (7\text{-}55)$$

- **For CO_2 systems.**

$$k_{ij} = 0.08058 - 0.77215k_{ij}^{\infty} - 1.8404\left(k_{ij}^{\infty}\right)^2 \qquad (7\text{-}56)$$

- **For H_2S systems.**

$$k_{ij} = 0.07654 + 0.017921k_{ij}^{\infty} \qquad (7\text{-}57)$$

- **For methane systems with compounds of 10 carbons or more.**

$$k_{ij} = 0.17985 - 2.6958k_{ij}^{\infty} - 10.853(k_{ij}^{\infty})^2 \qquad (7\text{-}58)$$

where

$$k_{ij}^{\infty} = \frac{-(\varepsilon_i - \varepsilon_j)^2}{2\varepsilon_i\varepsilon_j} \qquad (7\text{-}59)$$

and

$$\varepsilon_i = \frac{0.480453\sqrt{a_i}}{b_i} \qquad (7\text{-}60)$$

The two parameters a_i and b_i in Equation 7-60 were previously defined by Equations 7-28 and 7-29.

The major drawback in the SRK EOS is that the critical compressibility factor takes on the unrealistic universal critical compressibility of 0.333 for all substances. Consequently, the molar volumes are typically overestimated and, hence, densities are underestimated.

---

[2]An asymmetric mixture is defined as one in which two of the components are considerably different in their chemical behavior. Mixtures of methane with hydrocarbons of 10 or more carbon atoms can be considered asymmetric. Mixtures containing gases such as nitrogen or hydrogen are asymmetric.

Peneloux et al. (1982) [40] developed a procedure for improving the volumetric predictions of the SRK EOS by introducing a volume correction parameter c_i into the equation. This third parameter does not change the vapor-liquid equilibrium conditions determined by the unmodified SRK equation, i.e., the equilibrium ratio K_i, but it modifies the liquid and gas volumes. The proposed methodology, known as the *volume translation method*, uses the following expressions:

$$V^L_{corr} = V^L - \sum_i (x_i c_i) \qquad (7\text{-}61)$$

$$V^v_{corr} = V^v - \sum_i (y_i c_i) \qquad (7\text{-}62)$$

where V^L = uncorrected liquid molar volume, i.e., $V^L = Z^L RT/p$, ft^3/mol
V^v = uncorrected gas molar volume $V^v = Z^v RT/p$, ft^3/mol
V^L_{corr} = corrected liquid molar volume, ft^3/mol
V^v_{corr} = corrected gas molar volume, ft^3/mol
x_i = mole fraction of component i in the liquid phase
y_i = mole fraction of component i in the gas phase

The authors proposed six schemes for calculating the correction factor c_i for each component. For petroleum fluids and heavy hydrocarbons, Peneloux and coworkers suggested that the best correlating parameter for the correction factor c_i is the Rackett compressibility factor Z_{RA}. The correction factor is then defined mathematically by the following relationship:

$$c_i = 4.43797878(0.29441 - Z_{RA})T_{ci}/p_{ci} \qquad (7\text{-}63)$$

where c_i = correction factor for component i, $ft^3/lb\text{-}mol$
T_{ci} = critical temperature of component i, °R
p_{ci} = critical pressure of component i, psia

The parameter Z_{RA} is a unique constant for each compound. The values of Z_{RA} are in general not much different from those of the critical compressibility factors Z_c. If their values are not available, Peneloux et al. (1982) [40] proposed the following correlation for calculating c_i:

$$c_i = (0.0115831168 + 0.411844152\omega)\left(\frac{T_{ci}}{p_{ci}}\right) \qquad (7\text{-}64)$$

where ω_i = acentric factor of component i.

Example 7-7

Rework Example 7-6 by incorporating the Peneloux volume correction approach in the solution. Key information from Example 7-6 includes

- For gas: $Z^v = 0.9267$, $M_a = 20.89$.
- For liquid: $Z^L = 1.4121$, $M_a = 100.25$.
- $T = 160°F$, $p = 4000$ psi.

Solution

Step 1. Calculate the correction factor c_i using Equation 7-63:

Component	c_i	x_i	$c_i x_i$	y_i	$c_i y_i$
C_1	0.00839	0.45	0.003776	0.86	0.00722
C_2	0.03807	0.05	0.001903	0.05	0.00190
C_3	0.07729	0.05	0.003861	0.05	0.00386
C_4	0.1265	0.03	0.00379	0.02	0.00253
C_5	0.19897	0.01	0.001989	0.01	0.00198
C_6	0.2791	0.01	0.00279	0.005	0.00139
C_{7+}	0.91881	0.40	0.36752	0.005	0.00459
sum			0.38564		0.02349

Step 2. Calculate the uncorrected volume of the gas and liquid phase using the compressibility factors as calculated in Example 7-6:

$$V^v = \frac{(10.73)(620)(0.9267)}{4000} = 1.54119 \text{ ft}^3/\text{mol}$$

$$V^L = \frac{(10.73)(620)(1.4121)}{4000} = 2.3485 \text{ ft}^3/\text{mol}$$

Step 3. Calculate the corrected gas and liquid volumes by applying Equations 7-61 and 7-62:

$$V^L_{corr} = V^L - \sum_i (x_i c_i) = 2.3485 - 0.38564 = 1.962927 \text{ ft}^3/\text{mol}$$

$$V^v_{corr} = V^v - \sum_i (y_i c_i) = 1.54119 - 0.02349 = 1.5177 \text{ ft}^3/\text{mol}$$

Step 4. Calculate the corrected compressibility factors:

$$Z^v_{corr} = \frac{(4000)(1.5177)}{(10.73)(620)} = 0.91254$$

$$Z^L_{corr} = \frac{(4000)(1.962927)}{(10.73)(620)} = 1.18025$$

Step 5. Determine the corrected densities of both phases:

$$\rho = \frac{pM_a}{RTZ}$$

$$\rho^v = \frac{(4000)(20.89)}{(10.73)(620)(0.91254)} = 13.767 \text{ lb/ft}^3$$

$$\rho^L = \frac{(4000)(100.25)}{(10.73)(620)(1.18025)} = 51.07 \text{ lb/ft}^3$$

7.5 PENG-ROBINSON EQUATION OF STATE AND ITS MODIFICATIONS

Peng and Robinson (1976a) [41] conducted a comprehensive study to evaluate the use of the SRK equation of state for predicting the behavior of naturally occurring hydrocarbon systems. They illustrated the need for an improvement in the ability of the equation of state to predict liquid densities and other fluid properties, particularly in the vicinity of the critical region. As a basis for creating an improved model, Peng and Robinson proposed the following expression:

$$p = \frac{RT}{V - b} - \frac{a\alpha}{(V + b)^2 - cb^2}$$

where a, b, and α have the same significance as they have in the SRK model, and the parameter c is a whole number optimized by analyzing the values of the two terms Z_c and b/V_c as obtained from the equation. It is generally accepted that Z_c should be close to 0.28 and that b/V_c should be approximately 0.26. An optimized value of c = 2 gave Z_c = 0.307 and $(b/V_c) = 0.253$. Based on this value of c, Peng and Robinson proposed the following equation of state:

$$p = \frac{RT}{V - b} - \frac{a\alpha}{V(V + b) + b(V - b)} \tag{7-65}$$

Imposing the classical critical point conditions (Equation 7-3) on Equation 7-65 and solving for parameters a and b yields

$$a = \Omega_a \frac{R^2 T_c^2}{P_c} \tag{7-66}$$

$$b = \Omega_b \frac{RT_c}{P_c} \tag{7-67}$$

where $\Omega_a = 0.45724$ and $\Omega_b = 0.07780$. This equation predicts a universal critical gas compressibility factor Z_c of 0.307 compared to 0.333 for the

SRK model. Peng and Robinson also adopted Soave's approach for calculating the temperature-dependent parameter α:

$$\alpha = \left[1 + m\left(1 - \sqrt{T_r}\right)\right]^2 \qquad (7\text{-}68)$$

where

$$m = 0.3796 + 1.54226\omega - 0.2699\omega^2$$

Peng and Robinson (1978) proposed the following modified expression for m that is recommended for heavier components with acentric values $\omega > 0.49$:

$$m = 0.379642 + 1.48503\omega - 0.1644\omega^2 + 0.016667\omega^3 \qquad (7\text{-}69)$$

Rearranging Equation 7-65 into the compressibility factor form gives

$$Z^3 + (B - 1)Z^2 + (A - 3B^2 - 2B)Z - (AB - B^2 - B^3) = 0 \qquad (7\text{-}70)$$

where A and B are given by Equations 7-36 and 7-37 for pure components and by Equations 7-40 and 7-41 for mixtures.

Example 7-8

Using the composition given in Example 7-6, calculate the density of the gas phase and liquid phase using the Peng-Robinson EOS. Assume $k_{ij} = 0$.

Solution

Step 1. Calculate the mixture parameters $(a\alpha)_m$ and b_m for the gas and liquid phase, to give

- **For the gas phase.**

$$(a\alpha)_m = \sum_i \sum_j \left[y_i y_j \sqrt{a_i a_j \alpha_i \alpha_j} (1 - k_{ij})\right] = 10,423.54$$

$$b_m = \sum_i (y_i b_i) = 0.862528$$

- **For the liquid phase.**

$$(a\alpha)_m = \sum_i \sum_j \left[x_i x_j \sqrt{a_i a_j \alpha_i \alpha_j} (1 - k_{ij})\right] = 107,325.4$$

$$b_m = \sum_i (y_i b_i) = 1.69543$$

Step 2. Calculate the coefficients A and B, to give

- **For the gas phase.**

$$A = \frac{(a\alpha)_m P}{R^2 T^2} = \frac{(10,423.54)(4000)}{(10.73)^2 (620)^2} = 0.94209$$

$$B = \frac{b_m P}{RT} = \frac{(0.862528)(4000)}{(10.73)(620)} = 0.30669$$

- **For the liquid phase.**

$$A = \frac{(a\alpha)_m P}{R^2 T^2} = \frac{(107,325.4)(4000)}{(10.73)^2 (620)^2} = 9.700183$$

$$B = \frac{b_m P}{RT} = \frac{(1.636543)(4000)}{(10.73)(620)} = 1.020078$$

Step 3. Solve Equation 7-70 for the compressibility factor of the gas phase and the liquid phase to give

$$Z^3 + (B - 1)Z^2 + (A - 3B^2 - 2B)Z - (AB - B^2 - B^3) = 0$$

- **For the gas phase.** Substituting for A = 0.94209 and B = 0.30669 in the above equation gives

$$Z^V = 0.8625$$

- **For the liquid phase.** Substituting for A = 9.700183 and B = 1.020078 in the above equation gives

$$Z^L = 1.2645$$

Step 4. Calculate the density of both phases:

$$\rho^v = \frac{(4,000)(20.89)}{(10.73)(620)(0.8625)} = 14.566 \text{ lb/ft}^3$$

$$\rho^L = \frac{(4,000)(100.25)}{(10.73)(620)(1.2645)} = 47.67 \text{ lb/ft}^3$$

Applying the thermodynamic relationship, as given by Equation 7-43, to Equation 7-66 yields the following expression for the fugacity of a pure component:

$$\ln\left(\frac{f}{p}\right) = \ln(\Phi) = Z - 1 - \ln(Z - B) - \left[\frac{A}{2\sqrt{2}B}\right] \ln\left[\frac{Z + (1 + \sqrt{2})B}{Z + (1 - \sqrt{2})B}\right]$$

(7-71)

The fugacity coefficient of component i in a hydrocarbon liquid mixture is calculated from the following expression:

$$\ln\left(\frac{f_i^L}{x_i p}\right) = \ln(\Phi_i^L) = \frac{b_i(Z^L - 1)}{b_m} - \ln(Z^L - B)$$
$$- \left[\frac{A}{2\sqrt{2}B}\right]\left[\frac{2\Psi_i}{(a\alpha)_m} - \frac{b_i}{b_m}\right] \ln\left[\frac{Z^L + (1 + \sqrt{2})B}{Z^L - (1 - \sqrt{2})B}\right]$$

(7-72)

where the mixture parameters b_m, B, A, Ψ_i, and $(a\alpha)_m$ are as defined previously.

Equation 7-72 is also used to determine the fugacity coefficient of any component in the gas phase by replacing the composition of the liquid phase x_i with the composition of the gas phase y_i in calculating the composition-dependent terms of the equation, or

$$\ln\left(\frac{f^v}{y_i p}\right) = \ln(\Phi_i^v) = \frac{b_i(Z^v - 1)}{b_m} - \ln(Z^v - B)$$
$$- \left[\frac{A}{2\sqrt{2}B}\right]\left[\frac{2\Psi_i}{(a\alpha)_m} - \frac{b_i}{b_m}\right] \ln\left[\frac{Z^v + (1 + \sqrt{2})B}{Z^v - (1 - \sqrt{2})B}\right]$$

The set of binary interaction coefficients k_{ij} in Table 7-1 is traditionally used when predicting the volumetric behavior of a hydrocarbon mixture with the Peng and Robinson (PR) equation of state.

To improve the predictive capability of the PR EOS when describing mixtures containing N_2, CO_2, and CH_4, Nikos et al. (1986) [37] proposed a generalized correlation for generating the binary interaction coefficient k_{ij}. The authors correlated these coefficients with system pressure, temperature, and the acentric factor. These generalized correlations were originated with all the binary experimental data available in the literature. The authors proposed the following generalized form for k_{ij}:

$$k_{ij} = \delta_2 T_{rj}^2 + \delta_1 T_{rj} + \delta_0$$

(7-73)

where i refers to the principal components N_2, CO_2, or CH_4 and j refers to the other hydrocarbon components of the binary. The acentric factor-dependent coefficients δ_0, δ_1, and δ_2 are determined for each set of binaries by applying the following expressions:

TABLE 7-1 Binary Interaction Coefficients* k_{ij} for the peng and Robinson EOS

	CO_2	N_2	H_2S	C_1	C_2	C_3	$i-C_4$	$n-C_4$	$i-C_5$	$n-C_5$	C_6	C_7	C_8	C_9	C_{10}
CO_2	0														
N_2	0	0													
H_2S	0.135	0.130	0												
C_1	0.105	0.025	0.070	0											
C_2	0.130	0.010	0.085	0.005	0										
C_3	0.125	0.090	0.080	0.010	0.005	0									
$i-C_4$	0.120	0.095	0.075	0.035	0.005	0.005	0								
$n-C_4$	0.115	0.095	0.075	0.025	0.010	0.005	0.000	0							
$i-C_5$	0.115	0.100	0.070	0.050	0.020	0.005	0.005	0.005	0						
$n-C_5$	0.115	0.100	0.070	0.030	0.020	0.005	0.005	0.005	0.000	0					
C_6	0.115	0.110	0.070	0.030	0.020	0.010	0.005	0.000	0.000	0.000	0				
C_7	0.115	0.115	0.060	0.035	0.020	0.005	0.005	0.000	0.000	0.000	0.000	0			
C_8	0.115	0.120	0.060	0.040	0.020	0.005	0.005	0.000	0.000	0.000	0.000	0.000	0		
C_9	0.115	0.120	0.060	0.040	0.020	0.005	0.005	0.000	0.000	0.000	0.000	0.000	0.000	0	
C_{10}	0.115	0.125	0.055	0.045	0.020	0.005	0.005	0.005	0.000	0.000	0.000	0.000	0.000	0.000	0

*Notice that $k_{ij} = k_{ji}$.

- **For nitrogen-hydrocarbons.**

$$\delta_0 = 0.1751787 - 0.7043 \log(\omega_j) - 0.862066 \left[\log(\omega_i)\right]^2 \qquad (7\text{-}74)$$

$$\delta_1 = -0.584474 + 1.328 \log(\omega_j) + 2.035767 \left[\log(\omega_i)\right]^2 \qquad (7\text{-}75)$$

and

$$\delta_2 = 2.257079 + 7.869765 \log(\omega_j) + 13.50466 \left[\log(\omega_i)\right]^2 \\ + 8.3864 \left[\log(\omega)\right]^3 \qquad (7\text{-}76)$$

They also suggested the following pressure correction:

$$k_{ij} = k_{ij}(1.04 - 4.2 \times 10^{-5}p) \qquad (7\text{-}77)$$

where p is the pressure in pounds per square inch.
- **For methane-hydrocarbons.**

$$\delta_0 = -0.01664 - 0.37283 \log(\omega_j) + 1.31757 \left[\log(\omega_i)\right]^2 \qquad (7\text{-}78)$$

$$\delta_1 = 0.48147 + 3.35342 \log(\omega_j) - 1.0783 \left[\log(\omega_i)\right]^2 \qquad (7\text{-}79)$$

and

$$\delta_2 = -0.4114 - 3.5072 \log(\omega_j) - 1.0783 \left[\log(\omega_i)\right]^2 \qquad (7\text{-}80)$$

- **For CO_2-hydrocarbons.**

$$\delta_0 = 0.4025636 + 0.1748927 \log(\omega_j) \qquad (7\text{-}81)$$

$$\delta_1 = -0.94812 - 0.6009864 \log(\omega_j) \qquad (7\text{-}82)$$

and

$$\delta_2 = 0.741843368 + 0.441775 \log(\omega_j) \qquad (7\text{-}83)$$

For the CO_2 interaction parameters, the following pressure correction is suggested:

$$k'_{ij} = k_{ij}(1.044269 - 4.375 \times 10^{-5}p) \qquad (7\text{-}84)$$

Stryjek and Vera (1986) [59] proposed an improvement in the reproduction of vapor pressures of pure components by the PR EOS in the reduced temperature range from 0.7 to 1.0 by replacing the m term in Equation 7-72 with the following expression:

$$m_0 = 0.378893 + 1.4897153 - 0.17131848\omega^2 + 0.0196554\omega^3 \qquad (7\text{-}85)$$

To reproduce vapor pressures at reduced temperatures below 0.7, Stryjek and Vera further modified the m parameter in the PR equation by introducing an adjustable parameter m_1 characteristic of each compound to Equation 7-72. They proposed the following generalized relationship for the parameter m:

$$m = m_0 + \left[m_1 \left(1 + \sqrt{T_r} \right) (0.7 - T_r) \right] \qquad (7\text{-}86)$$

where T_r = reduced temperature of the pure component
$\quad m_0$ = defined by Equation 7-84
$\quad m_1$ = adjustable parameter

For all components with a reduced temperature above 0.7, Stryjek and Vera recommended setting $m_1 = 0$. For components with a reduced temperature greater than 0.7, the optimum values of m_1 for compounds of industrial interest follow:

Parameter m_1 of Pure Compounds			
Compound	m_1	Compound	m_1
Nitrogen	0.01996	Nonane	0.04104
Carbon dioxide	0.04285	Decane	0.04510
Water	−0.06635	Undecane	0.02919
Methane	−0.00159	Dodecane	0.05426
Ethane	0.02669	Tridecane	0.04157
Propane	0.03136	Tetradecane	0.02686
Butane	0.03443	Pentadecane	0.01892
Pentane	0.03946	Hexadecane	0.02665
Hexane	0.05104	Heptadecane	0.04048
Heptane	0.04648	Octadecane	0.08291
Octane	0.04464		

Due to the totally empirical nature of the parameter m_1, Stryjek and Vera could not find a generalized correlation for m_1 in terms of pure component parameters. They pointed out that the values of m_1 just given should be used without changes.

Jhaveri and Youngren (1984) [23] pointed out that, when applying the Peng-Robinson equation of state to reservoir fluids, the error associated with the equation in the prediction of gas-phase Z factors ranged from 3 to 5%, and the error in the liquid density predictions ranged from 6 to 12%. Following the procedure proposed by Peneloux and coworkers (see the SRK EOS), Jhaveri and Youngren introduced the volume correction parameter c_i to the PR EOS. This third parameter has the same units as the second parameter b_i of the unmodified PR equation and is defined by the following relationship:

$$c_i = S_i b_i \qquad (7\text{-}87)$$

where S_i = dimensionless parameter, called the *shift parameter*
b_i = Peng-Robinson covolume as given by Equation 7-67

The volume correction parameter c_i does not change the vapor-liquid equilibrium conditions, i.e., equilibrium ratio K_i. The corrected hydrocarbon phase volumes are given by the following expressions:

$$V^L_{corr} = V^L - \sum_{i=1}(x_i\, c_i)$$

$$V^v_{corr} = V^v - \sum_{i=1}(y_i\, c_i)$$

where V^L, V^v = volumes of the liquid phase and gas phase as calculated by the unmodified PR EOS, ft^3/mol
V^L_{corr}, V^v_{corr} = corrected volumes of the liquid and gas phase

Whitson and Brule (2000) [64] point out that the volume translation (correction) concept can be applied to any two-constant cubic equation, thereby eliminating the volumetric deficiency associated with application of EOS. Whitson and Brule extended the work of Jhaveri and Youngren (1984) and proposed the following shift parameters for selected pure components:

Shift Parameters for the PR EOS and SRK EOS

Compound	PR EOS	SRK EOS
N_2	−0.1927	−0.0079
CO_2	−0.0817	0.0833
H_2S	−0.1288	0.0466
C_1	−0.1595	0.0234
C_2	−0.1134	0.0605
C_3	−0.0863	0.0825
$i - C_4$	−0.0844	0.0830
$n - C_4$	−0.0675	0.0975
$i - C_5$	−0.0608	0.1022
$n - C_5$	−0.0390	0.1209
$n - C_6$	−0.0080	0.1467
$n - C_7$	0.0033	0.1554
$n - C_8$	0.0314	0.1794
$n - C_9$	0.0408	0.1868
$n - C_{10}$	0.0655	0.2080

Jhaveri and Youngren (1984) proposed the following expression for calculating the shift parameter for the C_{7+}:

$$S = 1 - \frac{d}{(M)^e}$$

where M = molecular weight of the heptanes-plus fraction
d, e = positive correlation coefficients

The authors proposed that, in the absence of the experimental information needed for calculating e and d, the power coefficient e can be set equal to 0.2051 and the coefficient d adjusted to match the C_{7+} density, with the values of d ranging from 2.2 to 3.2. In general, the following values may be used for C_{7+} fractions:

Hydrocarbon Family	d	e
Paraffins	2.258	0.1823
Naphthenes	3.044	0.2324
Aromatics	2.516	0.2008

To use the Peng-Robinson equation of state to predict the phase and volumetric behavior of mixtures, one must be able to provide the critical pressure, the critical temperature, and the acentric factor for each component in the mixture. For pure compounds, the required properties are well defined and known. Nearly all naturally occurring petroleum fluids contain a quantity of heavy fractions that are not well defined. These heavy fractions are often lumped together as the heptanes-plus fraction. The problem of how to adequately characterize the C_{7+} fractions in terms of their critical properties and acentric factors has been long recognized in the petroleum industry. Changing the characterization of C_{7+} fractions present in even small amounts can have a profound effect on the PVT properties and the phase equilibria of a hydrocarbon system as predicted by the Peng-Robinson equation of state.

The usual approach for such situations is to "tune" the parameters in the EOS in an attempt to improve the accuracy of prediction. During the tuning process, the critical properties of the heptanes-plus fraction and the binary interaction coefficients are adjusted to obtain a reasonable match with experimental data available on the hydrocarbon mixture.

Recognizing that the inadequacy of the predictive capability of the PR EOS lies with the improper procedure for calculating the parameters a, b, and α of the equation for the C_{7+} fraction, Ahmed (1991) [1] devised an approach for determining these parameters from the following two readily measured physical properties of C_{7+}: molecular weight, M_{7+}, and specific gravity, γ_{7+}.

The approach is based on generating 49 density values for the C_{7+} by applying the Riazi-Daubert correlation. These values were subsequently subjected to 10 temperature and 10 pressure values in the range of 60 to 300°F and 14.7 to 7000 psia, respectively. The Peng-Robinson EOS was then applied to match the 4900 generated density values by optimizing the parameters a, b, and α using a nonlinear regression model. The optimized parameters for the heptanes-plus fraction are given by the following expressions:

For the parameter a of C_{7+},

$$\alpha = \left[1 + m\left(1 - \sqrt{\frac{520}{T}}\right)\right]^2 \tag{7-88}$$

with m defined by

$$m = \frac{D}{A_0 + A_1 D} + A_2 M_{7+} + A_3 M_{7+}^2 + \frac{A_4}{M_{7+}} + A_5 \gamma_{7+}$$
$$+ A_6 \gamma_{7+}^2 + \frac{A_7}{\gamma_{7+}} \tag{7-89}$$

with the parameter D defined by the ratio of the molecular weight to the specific gravity of the heptanes-plus fraction, or

$$D = \frac{M_{7+}}{\gamma_{7+}}$$

where M_{7+} = molecular weight of C_{7+}
γ_{7+} = specific gravity of C_{7+}
$A_0 - A_7$ = coefficients as given in Table 7-2

For the parameters a and b of C_{7+}, the following generalized correlation is proposed:

$$a \text{ or } b = \left[\sum_{i=0}^{3} (A_i D^i)\right] + \frac{A_4}{D}\left[\sum_{i=5}^{6} (A_i \gamma_{7+}^{i-4})\right] + \frac{A_7}{\gamma_{7+}} \tag{7-90}$$

The coefficients A_0 through A_7 are included in Table 7-2.

To further improve the predictive capability of the Peng-Robinson EOS, the author optimized coefficients a, b, and m for nitrogen, CO_2, and methane by matching 100 Z-factor values for each of these components. Using a nonlinear regression model, the optimized values given in Table 7.2 are recommended.

To provide the modified PR EOS with a consistent procedure for determining the binary interaction coefficient k_{ij}, the following computational steps are proposed.

Step 1. Calculate the binary interaction coefficient between methane and the heptanes-plus fraction from

$$k_{c_1 - c_{7+}} = 0.00189T - 1.167059$$

where the temperature T is in °R.

TABLE 7-2   Coefficients for Equations 7-89 and 7-90

Coefficient	a	b	m
A_0	-2.433525×10^7	-6.8453198	-36.91776
A_1	8.3201587×10^3	1.730243×10^{-2}	$-5.2393763 \times 10^{-2}$
A_2	-0.18444102×10^2	$-6.2055064 \times 10^{-6}$	1.7316235×10^{-2}
A_3	3.6003101×10^{-2}	9.0910383×10^{-9}	$-1.3743308 \times 10^{-5}$
A_4	3.4992796×10^7	13.378898	12.718844
A_5	2.838756×10^7	7.9492922	10.246122
A_6	-1.1325365×10^7	-3.1779077	-7.6697942
A_7	6.418828×10^6	1.7190311	-2.6078099

Component	a	b	m in Eq. 7-88
CO_2	1.499914×10^4	0.41503575	-0.73605717
N_2	4.5693589×10^3	0.4682582	-0.97962859
C_1	7.709708×10^3	0.46749727	-0.549765

Step 2. Set

$$k_{CO_2-N_2} \qquad = 0.12$$
$$k_{CO_2-\text{hydrocarbon}} = 0.10$$
$$k_{N_2-\text{hydrocarbon}} = 0.10$$

Step 3. Adopting the procedure recommended by Petersen (1989), calculate the binary interaction coefficients between components heavier than methane (e.g., C_2, C_3) and the heptanes-plus fraction from

$$k_{C_n-C_{7+}} = 0.8 k_{C_{(n-1)}-C_{7+}}$$

where n is the number of carbon atoms of component C_n; e.g., the binary interaction coefficient between C_2 and C_{7+} is

$$k_{C_2-C_{7+}} = 0.8 k_{C_1-C_{7+}}$$

and the binary interaction coefficient between C_3 and C_{7+} is

$$k_{C_3-C_{7+}} = 0.8 k_{C_2-C_{7+}}$$

Step 4. Determine the remaining k_{ij} from

$$k_{ij} = k_{i-C_{7+}} \left[\frac{(M_j)^5 - (M_i)^5}{(M_{C_{7+}})^5 - (M_i)^5} \right]$$

where M is the molecular weight of any specified component. For example, the binary interaction coefficient between propane C_3 and butane C_4 is

$$k_{C_3-C_4} = k_{C_3-C_{7+}} \left[\frac{(M_{C_4})^5 - (M_{C_3})^5}{(M_{C_{7+}})^5 - (M_{C_3})^5} \right]$$

Applications of the Equation of State in Petroleum Engineering

8.1 DETERMINATION OF THE EQUILIBRIUM RATIOS

A flow diagram is presented in Figure 8-1 to illustrate the procedure of determining equilibrium ratios of a hydrocarbon mixture. For this type of calculation, the system temperature T, the system pressure p, and the overall composition of the mixture z_i must be known. The procedure is summarized in the following steps in conjunction with Figure 8-1.

Step 1. Assume an initial value of the equilibrium ratio for each component in the mixture at the specified system pressure and temperature. Wilson's equation can provide the starting K_i values:

$$K_i^A = \frac{p_{ci}}{p} \exp[5.37(1 + \omega_i)(1 - T_{ci}/T)]$$

where K_i^A = assumed equilibrium ratio of component i.

Step 2. Using the overall composition and the assumed K values, perform flash calculations to determine x_i, y_i, n_L, and n_v.

Step 3. Using the calculated composition of the liquid phase x_i, determine the fugacity coefficient Φ_i^L for each component in the liquid phase.

Step 4. Repeat step 3 using the calculated composition of the gas phase y_i to determine Φ_i^v.

Step 5. Calculate the new set of equilibrium ratios from

$$K_i = \frac{\Phi_i^L}{\Phi_i^v}$$

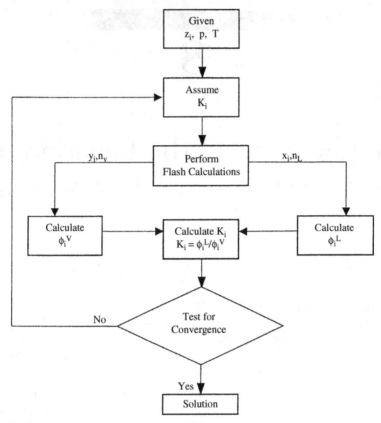

FIGURE 8-1   Flow diagram of the equilibrium ratio determination by an equation of state.

Step 6. Check for the solution by applying the following constraint:

$$\sum_{i=1}^{n} [K_i/K_i^A - 1]^2 \leq \varepsilon$$

where ε = preset error tolerance, e.g., 0.0001
       n = number of components in the system

If these conditions are satisfied, then the solution has been reached. If not, steps 1 through 6 are repeated using the calculated equilibrium ratios as initial values.

8.2 DETERMINATION OF THE DEW-POINT PRESSURE

A saturated vapor exists for a given temperature at the pressure at which an infinitesimal amount of liquid first appears. This pressure is referred to as the *dew-point pressure* p_d. The dew-point pressure of a mixture is described mathematically by the following two conditions:

$$y_i = z_i \quad 1 \le i \le n$$
$$n_v = 1 \tag{8-1}$$

and

$$\sum_{i=1}^{n} \left[\frac{z_i}{K_i} \right] = 1 \tag{8-2}$$

Applying the definition of K_i in terms of the fugacity coefficient to Equation 8-2 gives

$$\sum_{i=1}^{n} \left[\frac{z_i}{k_i} \right] = \sum_{i=1}^{n} \left[\frac{z_i}{(\Phi_i^L/\Phi_i^v)} \right] = \sum \left[\left(\frac{z_i}{\Phi_i^L} \right) \frac{f_i^v}{z_i p_d} \right] = 1$$

or

$$p_d = \sum_{i=1}^{n} \left[\frac{f_i^v}{\Phi_i^L} \right]$$

This equation is arranged to give

$$f(p_d) = \sum_{i=1}^{n} \left[\frac{f_i^v}{\Phi_i^L} \right] - p_d = 0 \tag{8-3}$$

where p_d = dew-point pressure, psia

f_i^v = fugacity of component i in the vapor phase, psia

Φ_i^L = fugacity coefficient of component i in the liquid phase

Equation 8-3 can be solved for the dew-point pressure by using the Newton–Raphson iterative method. To use the interative method, the derivative of Equation 8-3 with respect to the dew-point pressure p_d is required. This derivative is given by the following expression:

$$\frac{\partial f}{\partial p_d} = \sum_{i=1}^{n} \left[\frac{\Phi_i^L(\partial f_i^v/\partial p_d) - f_i^v(\partial \Phi_i^L/\partial p_d)}{(\Phi_i^L)^2} \right] - 1 \tag{8-4}$$

The two derivatives in the preceding equation can be approximated numerically as follows:

$$\frac{\partial f^v}{\partial p_d} = \left[\frac{f_i^v(p_d + \Delta p_d) - f_i^v(p_d - \Delta p_d)}{2\Delta p_d}\right] \tag{8-5}$$

and

$$\frac{\partial f_i^L}{\partial p_d} = \left[\frac{\Phi_i^L(p_d + \Delta p_d) - \Phi_i^L(p_d - \Delta p_d)}{2\Delta p_d}\right] \tag{8-6}$$

where Δp_d = pressure increment, 5 psia, for example
$f_i^v(p_d + \Delta p_d)$ = fugacity of component i at $(p_d + \Delta p_d)$
$f_i^v(p_d - \Delta p_d)$ = fugacity of component i at $(p_d - \Delta p_d)$
$\Phi_i^L(p_d + \Delta p_d)$ = fugacity coefficient of component i at $(p_d + \Delta p_d)$
$\Phi_i^L(p_d - \Delta p_d)$ = fugacity coefficient of component i at $(p_d - \Delta p_d)$
Φ_i^L = fugacity coefficient of component i at p_d

The computational procedure of determining p_d is summarized in the following steps:

Step 1. Assume an initial value for the dew-point pressure p_d^A.
Step 2. Using the assumed value of p_d^A, calculate a set of equilibrium ratios for the mixture using any of the previous correlations, e.g., Wilson's correlation.
Step 3. Calculate the composition of the liquid phase, i.e., composition of the droplets of liquid, by applying the mathematical definition of K_i, to give

$$x_i = \frac{z_i}{K_i}$$

Note that $y_i = z_i$.
Step 4. Calculate f_i^v using the composition of the gas phase z_i, and Φ_i^L using the composition of liquid phase x_i at the following three pressures:

- p_d^A.
- $p_d^A + \Delta p_d$.
- $p_d^A - \Delta p_d$.

where p_d^A is the assumed dew-point pressure and Δp_d is a selected pressure increment of 5 to 10 psi.
Step 5. Evaluate the function $f(p_d)$, i.e., Equation 8-3, and its derivative by using Equations 8-4 through 8-6.
Step 6. Using the values of the function $f(p_d)$ and the derivative $\partial f/\partial p_d$ as determined in step 5, calculate a new dew-point pressure by applying the Newton-Raphson formula:

$$p_d = p_d^A - f(p_d)/[\partial f/\partial p_d] \tag{8-7}$$

Step 7. The calculated value of p_d is checked numerically against the assumed value by applying the following condition:

$$|p_d - p_d^A| \leq 5$$

If this condition is met, then the correct dew-point pressure p_d has been found. Otherwise, steps 3 through 6 are repeated by using the calculated p_d as the new value for the next iteration. A set of equilibrium ratios must be calculated at the new assumed dew-point pressure from

$$k_i = \frac{\Phi_i^L}{\Phi_i^V}$$

8.3 DETERMINATION OF THE BUBBLE-POINT PRESSURE

The bubble-point pressure p_b is defined as the pressure at which the first bubble of gas is formed. Accordingly, the bubble-point pressure is defined mathematically by the following equations:

$$\begin{aligned} x_i &= z_i \quad 1 \leq i \leq n \\ n_L &= 1.0 \end{aligned} \tag{8-8}$$

and

$$\sum_{i=1}^{n} [z_i K_i] = 1 \tag{8-9}$$

Introducing the concept of the fugacity coefficient into Equation 8-9 gives

$$\sum_{i=1}^{n} \left[z_i \frac{\Phi_i^L}{\Phi_i^V} \right] = \sum_{i=1}^{n} \left[z_i \frac{\left(\frac{f_i^L}{z_i p_b} \right)}{\Phi_i^V} \right] = 1$$

Rearranging,

$$p_b = \sum_{i=1}^{n} \left[\frac{f_i^L}{\Phi_i^V} \right]$$

or

$$f(p_b) = \sum_{i=1}^{n} \left[\frac{f_i^L}{\Phi_i^V} \right] - p_b = 0 \tag{8-10}$$

The iteration sequence for calculation of p_b from this function is similar to that of the dew-point pressure, which requires differentiating the function with respect to the bubble-point pressure, or

$$\frac{\partial f}{\partial p_b} = \sum_{i=1}^{n} \left[\frac{\Phi_i^v(\partial f_i^L/\partial p_b) - f_i^L(\partial \Phi_i^v/\partial p_b)}{(\Phi_i^v)^2} \right] - 1 \tag{8-11}$$

8.4 THREE-PHASE EQUILIBRIUM CALCULATIONS

Two- and three-phase equilibria occur frequently during the processing of hydrocarbon and related systems. Peng and Robinson (1976b) [42] proposed a three-phase equilibrium calculation scheme of systems that exhibit a water-rich liquid phase, a hydrocarbon-rich liquid phase, and a vapor phase.

Applying the principle of mass conservation to 1 mol of a water-hydrocarbon in a three-phase state of thermodynamic equilibrium at a fixed temperature T and pressure p gives

$$n_L + n_w + n_v = 1 \tag{8-12}$$

$$n_L x_i + n_w x_{wi} + n_v y_i = z_i \tag{8-13}$$

$$\sum_{i}^{n} x_i = \sum_{i=1}^{n} x_{wi} = \sum_{i=1}^{n} y_i = \sum_{i=1}^{n} z_i = 1 \tag{8-14}$$

where n_L, n_w, n_v = number of moles of the hydrocarbon-rich liquid, the water-rich liquid, and the vapor, respectively

x_i, x_{wi}, y_i = mole fraction of component i in the hydrocarbon-rich liquid, the water-rich liquid, and the vapor, respectively.

The equilibrium relations between the compositions of each phase are defined by the following expressions:

$$K_i = \frac{y_i}{x_i} = \frac{\Phi_i^L}{\Phi_i^v} \tag{8-15}$$

and

$$K_{wi} = \frac{y_i}{x_{wi}} = \frac{\Phi_i^w}{\Phi_i^v} \tag{8-16}$$

where K_i = equilibrium ratio of component i between the vapor and hydrocarbon-rich liquid

K_{wi} = equilibrium ratio of component i between the vapor and water-rich liquid

Φ_i^L = fugacity coefficient of component i in the hydrocarbon-rich liquid

Φ_i^v = fugacity coefficient of component i in the vapor phase

Φ_i^w = fugacity coefficient of component i in the water-rich liquid

Combining Equations 8-12 through 8-16 gives the following conventional nonlinear equations:

$$\sum_{i=1} x_i = \sum_{i=1} \left[\frac{z_i}{n_L(1-K_i) + n_w\left(\dfrac{K_i}{K_{wi}} - K_i\right) + K_i} \right] = 1 \qquad (8\text{-}17)$$

$$\sum_{i=1} x_{wi} = \sum_{i=1} \left[\frac{z_i(K_i/K_{wi})}{n_L(1-K_i) + n_w\left(\dfrac{K_i}{K_{wi}} - K_i\right) + K_i} \right] = 1 \qquad (8\text{-}18)$$

$$\sum_{i=1} y_i = \sum_{i=1} \left[\frac{z_i K_i}{n_L(1-K_i) + n_w\left(\dfrac{K_i}{K_{wi}} - K_i\right) + K_i} \right] = 1 \qquad (8\text{-}19)$$

Assuming that the equilibrium ratios between phases can be calculated, these equations are combined to solve for the two unknowns n_L and n_v, and hence x_i, x_{wi}, and y_i. It is the nature of the specific equilibrium calculation that determines the appropriate combination of Equations 8-17 through 8-19. The combination of these three expressions can then be used to determine the phase and volumetric properties of the three-phase system.

There are essentially three types of phase behavior calculations for the three-phase system:

1. Bubble-point prediction.
2. Dew-point prediction.
3. Flash calculation.

Peng and Robinson (1980) [43] proposed the following combination schemes of Equations 8-17 through 8-19.

- **For the bubble-point pressure determination.**

$$\sum_i x_i - \sum_i x_{wi} = 0 \quad \left[\sum_i y_i \right] - 1 = 0$$

Substituting Equations 8-17 through 8-19 in this relationship gives

$$f(n_L, n_w) = \sum_i \left[\frac{z_i(1 - K_i/K_{wi})}{n_L(1 - K_i) + n_w(K_i/K_{wi} - K_i) + K_i} \right] = 0 \qquad (8\text{-}20)$$

and

$$g(n_L, n_w) = \sum_i \left[\frac{z_i K_i}{n_L(1 - K_i) + n_w(K_i/K_{wi} - K_i) + K_i} \right] - 1 = 0 \qquad (8\text{-}21)$$

- **For the dew-point pressure.**

$$\sum_i x_{wi} - \sum_i y_i = 0 \quad \left[\sum_i x_i \right] - 1 = 0$$

Combining with Equations 8-17 through 8-19 yields

$$f(n_L, n_w) = \sum_i \left[\frac{z_i K_i(1/K_{wi} - 1)}{n_L(1 - K_i) + n_w(K_i/K_{wi} - K_i) + K_i} \right] = 0 \qquad (8\text{-}22)$$

and

$$g(n_L, n_w) = \sum_i \left[\frac{z_i}{n_L(1 - K_i) + n_w(K_i/K_{wi} - K_i) + K_i} \right] - 1 = 0 \qquad (8\text{-}23)$$

- **For flash calculations.**

$$\sum_i x_i - \sum_i y_i = 0 \quad \left[\sum_i x_{wi} \right] - 1 = 0$$

or

$$f(n_L, n_w) = \sum_i \left[\frac{z_i(1 - K_i)}{n_L(1 - K_i) + n_w(K_i/K_{wi} - K_i) + K_i} \right] = 0 \qquad (8\text{-}24)$$

and

$$g(n_L, n_w) = \sum_i \left[\frac{z_i K_i/K_{wi}}{n_L(1 - K_i) + n_w(K_i/K_{wi} - K_i) + K_i} \right] - 1.0 = 0 \qquad (8\text{-}25)$$

Note that in performing any of these property predictions, we always have two unknown variables, n_L and n_w, and, between them, two equations. Providing that the equilibrium ratios and the overall composition are known, the equations can be solved simultaneously by using the appropriate iterative technique, e.g., the Newton-Raphson method. The application of this iterative technique for solving Equations 8-24 and 8-25 is summarized in the following steps.

Step 1. Assume initial values for the unknown variables n_L and n_w.

Step 2. Calculate new values of n_L and n_w by solving the following two linear equations:

$$
\begin{bmatrix} n_L \\ n_w \end{bmatrix}^{new} = \begin{bmatrix} n_L \\ n_w \end{bmatrix} - \begin{bmatrix} \partial f/\partial n_L & \partial f/\partial n_w \\ \partial g/\partial n_L & \partial g/\partial n_w \end{bmatrix}^{-1} \begin{bmatrix} f(n_L, n_w) \\ g(n_L, n_w) \end{bmatrix}
$$

where $f(n_L, n_w)$ = value of the function $f(n_L, n_w)$ as expressed by Equation 8-24

$g(n_L, n_w)$ = value of the function $g(n_L, n_w)$ as expressed by Equation 8-25

The first derivative of these functions with respect to n_L and n_w are given by the following expressions:

$$
(\partial f/\partial n_L) = \sum_{i=1} \left[\frac{-z_i(1 - K_i)^2}{[n_L(1 - K_i) + n_w(K_i/K_{wi} - K_i) + K_i]^2} \right]
$$

$$
(\partial f/\partial n_w) = \sum_{i=1} \left[\frac{-z_i(1 - K_i)(K_i/K_{wi} - K_i)}{[n_L(1 - K_i) + n_w(K_i/K_{wi} - K_i) + K_i]^2} \right]
$$

$$
(\partial g/\partial n_L) = \sum_{i=1} \left[\frac{-z_i(K_i/K_{wi})(1 - K_i)}{[n_L(1 - K_i) + n_w(K_i/K_{wi} - K_i) + K_i]^2} \right]
$$

$$
(\partial g/\partial n_w) = \sum_{i=1} \left[\frac{-z_i(K_i K_{wi})(K_i/K_{wi} - K_i)}{[n_L(1 - K_i) + n_w(K_i/K_{wi} - K_i) + K_i]^2} \right]
$$

Step 3. The new calculated values of n_L and n_w are then compared with the initial values. If no changes in the values are observed, then the correct values of n_L and n_w have been obtained. Otherwise, the preceding steps are repeated with the new calculated values used as initial values.

Peng and Robinson (1980) [43] proposed two modifications when using their equation of state for three-phase equilibrium calculations. The first modification concerns the use of the parameter α as expressed by Equation 7-68 for the water compound. Peng and Robinson suggested that, when the reduced temperature of this compound is less than 0.85, the following equation is applied:

$$
\alpha = [1.0085677 + 0.82154(1 - T_r^{0.5})]^2 \tag{8-26}
$$

where T_r is the reduced temperature $(T/T_c)H_2O$ of the water component.

The second modification concerns the application of Equation 7-38 for the water-rich liquid phase. A temperature-dependent binary interaction coefficient was introduced into the equation to give

$$(a\alpha)_m = \sum_i \sum_j [x_{wi} x_{wj} (a_i a_j \alpha_i \alpha_j)^{0.5} (1 - \tau_{ij})] \qquad (8\text{-}27)$$

where τ_{ij} is a temperature-dependent binary interaction coefficient. Peng and Robinson proposed graphical correlations for determining this parameter for each aqueous binary pair. Lim et al. (1984) [31] expressed these graphical correlations mathematically by the following generalized equation:

$$\tau_{ij} = a_1 \left[\frac{T}{T_{ci}}\right]^2 \left[\frac{p_{ci}}{p_{cj}}\right]^2 + a_2 \left[\frac{T}{T_{ci}}\right]\left[\frac{p_{ci}}{p_{cj}}\right] + a_3 \qquad (8\text{-}28)$$

where T = system temperature, °R
  T_{ci} = critical temperature of the component of interest, °R
  p_{ci} = critical pressure of the component of interest, psia
  p_{cj} = critical pressure of the water compound, psia

Values of the coefficients a_1, a_2, and a_3 of the preceding polynomial for selected binaries follow:

Component i	a_1	a_2	a_3
C_1	0	1.659	−0.761
C_2	0	2.109	−0.607
C_3	−18.032	9.441	−1.208
$n - C_4$	0	2.800	−0.488
$n - C_6$	49.472	−5.783	−0.152

For selected nonhydrocarbon components, values of interaction parameters are given by the following expressions:

- **For N_2-H_2O binary.**

$$\tau_{ij} = 0.402(T/T_{ci}) - 1.586 \qquad (8\text{-}29)$$

where τ_{ij} = binary parameter between nitrogen and the water compound
  T_{ci} = critical temperature of nitrogen, °R

- **For CO_2-H_2O binary**.

$$\tau_{ij} = -0.074 \left[\frac{T}{T_{ci}} \right]^2 + 0.478 \left[\frac{T}{T_{ci}} \right] - 0.503 \qquad (8\text{-}30)$$

where T_{ci} is the critical temperature of CO_2.

In the course of making phase equilibrium calculations, it is always desirable to provide initial values for the equilibrium ratios, so the iterative procedure can proceed as reliably and rapidly as possible. Peng and Robinson (1980) [43] adopted Wilson's equilibrium ratio correlation to provide initial K values for the hydrocarbon-vapor phase:

$$K_i = p_{ci}/p \exp[5.3727(1 + \omega_i)(1 - T_{ci}/T)]$$

For the water-vapor phase, Peng and Robinson proposed the following expression:

$$K_{wi} = 10^6 [p_{ci} T/(T_{ci} p)]$$

8.5 VAPOR PRESSURE FROM EQUATION OF STATE

The calculation of the vapor pressure of a pure component through an EOS is usually made by the same trial-and-error algorithms used to calculate vapor-liquid equilibria of mixtures. Soave (1972) [54] suggested that the van der Waals (vdW), Soave-Redlich-Kwong (SRK), and Peng-Robinson (PR) equations of state can be written in the following generalized form:

$$p = \frac{RT}{v - b} - \frac{a\alpha}{v^2 = \mu vb + wb^2} \qquad (8\text{-}31)$$

with

$$a = \Omega_a \frac{R^2 T_c^2}{P_c}$$

$$b = \Omega_b \frac{RT_c}{P_c}$$

where the values of u, w, Ω_a, and Ω_b for three different equations of state are as follows:

EOS	u	w	Ω_a	Ω_b
vdW	0	0	0.421875	0.125
SRK	1	0	0.42748	0.08664
PR	2	−1	0.45724	0.07780

Soave (1972) [54] introduced the reduced pressure p_r and reduced temperature T_r to these equations to give

$$A = \frac{a\alpha p}{R^2 T^2} = \Omega_a \frac{\alpha p_r}{T_r} \tag{8-32}$$

$$B = \frac{bp}{RT} = \Omega_b \frac{p_r}{T_r} \tag{8-33}$$

and

$$\frac{A}{B} = \frac{\Omega_a}{\Omega_b} \left(\frac{\alpha}{T_r} \right) \tag{8-34}$$

where

$$p_r = p/p_c$$
$$T_r = T/T_c$$

In the cubic form and in terms of the Z factor, these three equations of state can be written

$$\begin{aligned} \text{VdW: } & Z^3 - Z^2(1+B) + ZA - AB = 0 \\ \text{SRK: } & Z^3 - Z^2 + Z(A - B - B^2) - AB = 0 \\ \text{PR: } & Z^3 - Z^2(1-B) + Z(A - 3B^2 - 2B) - (AB - B^2 - B^3) = 0 \end{aligned} \tag{8-35}$$

and the pure component fugacity coefficient is given by

$$\text{VdW: } \ln(f/p) = Z - 1 - \ln(Z - B) - \frac{A}{Z}$$

$$\text{SRK: } \ln(f/p) = Z - 1 - \ln(Z - B) - \left(\frac{A}{B} \right) \ln \left(1 + \frac{B}{Z} \right)$$

$$\text{PR: } \ln(f/p) = Z - 1 - \ln(Z - B) - \left(\frac{A}{2\sqrt{2}B} \right) \ln \left(\frac{Z + (1 + \sqrt{2})B}{Z - (1 - \sqrt{2})B} \right)$$

A typical iterative procedure for the calculation of pure component vapor pressure at any temperature T through one of the EOSs is summarized next:

Step 1. Calculate the reduced temperature, i.e., $T_r = T/T_c$.
Step 2. Calculate the ratio A/B from Equation 8-34.
Step 3. Assume a value for B.
Step 4. Solve Equation 8-35 and obtain Z^L and Z^v, i.e., smallest and largest roots, for both phases.
Step 5. Substitute Z^L and Z^v into the pure component fugacity coefficient and obtain $\ln(f/p)$ for both phases.

Step 6. Compare the two values of f/p. If the isofugacity condition is not satisfied, assume a new value of B and repeat steps 3 through 6.

Step 7. From the final value of B, obtain the vapor pressure from Equation 8-33, or

$$B = \Omega_b \frac{(p_v/p_c)}{T_r}$$

Solving for p_v gives

$$p_v = \frac{B T_r P_c}{\Omega_b}$$

Splitting and Lumping Schemes of the Plus Fraction

The hydrocarbon-plus fractions that constitute a significant portion of naturally occurring hydrocarbon fluids create major problems when predicting the thermodynamic properties and the volumetric behavior of these fluids by equations of state. These problems arise due to the difficulty of properly characterizing the plus fractions (heavy ends) in terms of their critical properties and acentric factors.

Whitson (1980) [63] and Maddox and Erbar (1982, 1984) [33, 34], among others, have shown the distinct effect of the heavy fractions characterization procedure on PVT relationship prediction by equations of state. Usually, these undefined plus fractions, commonly known as the C_{7+} fractions, contain an undefined number of components with a carbon number higher than 6. Molecular weight and specific gravity of the C_{7+} fraction may be the only measured data available.

In the absence of detailed analytical data for the plus fraction in a hydrocarbon mixture, erroneous predictions and conclusions can result if the plus fraction is used directly as a single component in the mixture phase behavior calculations. Numerous authors have indicated that these errors can be substantially reduced by "splitting" or "breaking down" the plus fraction into a manageable number of fractions (pseudo-components) for equation of state calculations.

The problem, then, is how to adequately split a C_{7+} fraction into a number of pseudo-components characterized by

- Mole fractions.
- Molecular weights.
- Specific gravities.

These characterization properties, when properly M_{7+} combined, should match the measured plus fraction properties, i.e., $(M)_{7+}$ and $(\gamma)_{7+}$.

9.1 SPLITTING SCHEMES

Splitting schemes refer to the procedures of dividing the heptanes-plus fraction into hydrocarbon groups with a single carbon number (C_7, C_8, C_9, etc.) and are described by the same physical properties used for pure components.

Several authors have proposed different schemes for extending the molar distribution behavior of C_{7+}, i.e., the molecular weight and specific gravity. In general, the proposed schemes are based on the observation that lighter systems such as condensates usually exhibit exponential molar distribution, while heavier systems often show left-skewed distributions. This behavior is shown schematically in Figure 9-1.

Three important requirements should be satisfied when applying any of the proposed splitting models:

1. The sum of the mole fractions of the individual pseudo-components is equal to the mole fraction of C_{7+}.
2. The sum of the products of the mole fraction and the molecular weight of the individual pseudo-components is equal to the product of the mole fraction and molecular weight of C_{7+}.
3. The sum of the product of the mole fraction and molecular weight divided by the specific gravity of each individual component is equal to that of C_{7+}.

These requirements can be expressed mathematically by the following relationship:

$$\sum_{n=7}^{N+} z_n = z_{7+} \qquad (9\text{-}1)$$

FIGURE 9-1 Exponential and left-skewed distribution functions.

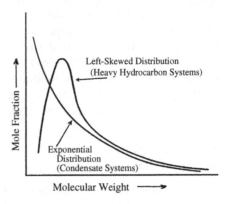

$$\sum_{n=7}^{N+}[z_n M_n] = z_{7+}M_{7+} \tag{9-2}$$

$$\sum_{n=7}^{N+}\frac{z_n M_n}{\gamma_n} = \frac{z_{7+}M_{7+}}{\gamma_{7+}} \tag{9-3}$$

where z_{7+} = mole fraction of C_{7+}
  n = number of carbon atoms
  N_+ = last hydrocarbon group in C_{7+} with n carbon atoms, e.g., 20+
  z_n = mole fraction of the psuedo-component with n carbon atoms
  M_{7+}, γ_{7+} = measure of molecular weight and specific gravity of C_{7+}
  M_n, γ_n = molecular weight and specific gravity of the pseudo-component with n carbon atoms

Several splitting schemes have been proposed. These schemes, as discussed next, are used to predict the compositional distribution of the heavy plus fraction.

Katz's Method

Katz (1983) presented an easy-to-use graphical correlation for breaking down into pseudo-components the C_{7+} fraction present in condensate systems. The method was originated by studying the compositional behavior of six condensate systems using detailed extended analysis. On a semilog scale, the mole percent of each constituent of the C_{7+} fraction versus the carbon number in the fraction was plotted. The resulting relationship can be conveniently expressed mathematically by the following expression:

$$z_n = 1.38205 z_{7+} e^{-0.25903n} \tag{9-4}$$

where z_{7+} = mole fraction of C_{7+} in the condensate system
  n = number of carbon atoms of the pseudo-component
  z_n = mole fraction of the pseudo-component with number of carbon atoms of n

Equation 9-4 is repeatedly applied until Equation 9-1 is satisfied. The molecular weight and specific gravity of the last pseudo-component can be calculated from Equations 9-2 and 9-3, respectively.

The computational procedure of Katz's method is best explained through the following example.

Example 9-1

A naturally occurring condensate gas system has the following composition:

Component	z_i
C_1	0.9135
C_2	0.0403
C_3	0.0153
$i - C_4$	0.0039
$n - C_4$	0.0043
$i - C_5$	0.0015
$n - C_5$	0.0019
C_6	0.0039
C_{7+}	0.0154

The molecular weight and specific gravity of C_{7+} are 141.25 and 0.797, respectively.

a. Using Katz's splitting scheme, extend the compositional distribution of C_{7+} to the pseudo-fraction C_{16+}.
b. Calculate M, γ, T_b, p_c, T_c, and ω of C_{16+}.

Solution

a. Extend the Computational Distribution

Applying Equation 9-4 with $z_{7+} = 0.0154$ gives

n	Experimental z_n	Equation 9-4 z_n
7	0.00361	0.00347
8	0.00285	0.00268
9	0.00222	0.00207
10	0.00158	0.001596
11	0.00121	0.00123
12	0.00097	0.00095
13	0.00083	0.00073
14	0.00069	0.000566
15	0.00050	0.000437
16+	0.00094	0.001671*

*This value is obtained by applying Equations 9-1,

i.e., $0.0154 - \sum_{n=7}^{15} z_n = 0.001671$.

b. Calculate M, γ, T_b, p_c, T_c and ω

Step 1. Calculate the molecular weight and specific gravity of C_{16+} by solving Equations 9-2 and 9-3 for these properties:

$$M_{16+} = z_{7+}M_{7+} - \left[\left(\frac{1}{z_{16+}}\right)\sum_{n=7}^{15}(z_n \cdot M_n)\right]$$

and

$$\gamma_{16+} = \frac{z_{16+}M_{16+}}{(z_{7+}M_{7+}/\gamma_{7+})} - \sum_{n=7}^{15}\left(\frac{z_n M_n}{\gamma_n}\right)$$

where M_n, γ_n = molecular weight and specific gravity of the hydrocarbon group with n carbon atoms. The calculations are performed in the following tabulated form:

n	z_n	M_n	$z_n M_n$	γ_n	$z_n \cdot M/\gamma_n$
7	0.00347	96	0.33312	0.727	0.4582
8	0.00268	107	0.28676	0.749	0.3829
9	0.00207	121	0.25047	0.768	0.3261
10	0.001596	134	0.213864	0.782	0.27348
11	0.00123	147	0.18081	0.793	0.22801
12	0.00095	161	0.15295	0.804	0.19024
13	0.00073	175	0.12775	0.815	0.15675
14	0.000566	190	0.10754	0.826	0.13019
15	0.000437	206	0.09002	0.836	0.10768
16+	0.001671	—	—	—	—
			1.743284		2.25355

$$M_{16+} = \frac{(0.0154)(141.25) - 1.743284}{0.001671} = 258.5$$

$$\gamma_{16+} = \frac{(0.001671)(258.5)}{\dfrac{(0.0154)(141.25)}{(0.797)}} - 2.25355 = 0.908$$

Step 2. Calculate the boiling points, critical pressure, and critical temperature of C_{16+} using the Riazi-Daubert correlation to give

$$T_b = 1{,}136°R$$
$$p_c = 215 \text{ psia}$$
$$T_c = 1{,}473°R$$

Step 3. Calculate the acentric factor of C_{16+} by applying the Edmister correlation to give $\omega = 0.684$.

Lohrenz's Method

Lohrenz et al. (1964) [32] proposed that the heptanes-plus fraction could be divided into pseudo-components with carbon numbers ranging from 7 to 40. They mathematically stated that the mole fraction z_n is related to its number of carbon atoms n and the mole fraction of the hexane fraction z_6 by the expression

$$z_n = z_6 e^{A(n-6)^2 + B(n-6)} \tag{9-5}$$

The constants A and B are determined such that the constraints given by Equations 9-1 through 9-3 are satisfied.

The use of Equation 9-5 assumes that the individual C_{7+} components are distributed through the hexane mole fraction and tail off to an extremely small quantity of heavy hydrocarbons.

Example 9-2

Rework Example 9-1 using the Lohrenz splitting scheme and assuming that a partial molar distribution of C_{7+} is available. The composition is as follows:

Component	z_i
C_1	0.9135
C_2	0.0403
C_3	0.0153
$i - C_4$	0.0039
$n - C_4$	0.0043
$i - C_5$	0.0015
$n - C_5$	0.0019
C_6	0.0039
C_7	0.00361
C_8	0.00285
C_9	0.00222
C_{10}	0.00158
C_{11+}	0.00514

Solution

Step 1. Determine the coefficients A and B of Equation 9-5 by the least-squares fit to the mole fractions C_6 through C_{10} to give $A = 0.03453$ and $B = 0.08777$.

Step 2. Solve for the mole fraction of C_{10} through C_{15} by applying Equation 9-5 and setting $z_6 = 0.0039$:

Component	Experimental z_n	Equation 9.5 z_n
C_7	0.00361	0.00361
C_8	0.00285	0.00285
C_9	0.00222	0.00222
C_{10}	0.00158	0.00158
C_{11}	0.00121	0.00106
C_{12}	0.00097	0.00066
C_{13}	0.00083	0.00039
C_{14}	0.00069	0.00021
C_{15}	0.00050	0.00011
C_{16+}	0.00094	0.00271*

*Obtained by applying Equation 9-1.

Step 3. Calculate the molecular weight and specific gravity of C_{16+} by applying Equations 9-2 and 9-3 to give $(M)_{16+} = 233.3$ and $(\gamma)_{16+} = 0.943$.

Step 4. Solve for T_b, p_c, T_c, and ω by applying the Riazi-Daubert and Edmister correlations to give

$$T_b = 1,103°R$$
$$p_c = 251 \text{ psia}$$
$$T_c = 1,467°R$$
$$\omega = 0.600$$

Pedersen's Method

Pedersen et al. (1982) [39] proposed that, for naturally occurring hydrocarbon mixtures, an exponential relationship exists between the mole fraction of a component and the corresponding carbon number. They expressed this relationship mathematically in the following form:

$$z_n = e^{(n-A)/B} \tag{9-6}$$

where A and B are constants.

For condensates and volatile oils, Pedersen and coworkers suggested that A and B can be determined by a least-squares fit to the molar distribution of the lighter fractions. Equation 9-6 can then be used to calculate the molar content of each of the heavier fractions by extrapolation. The classical constraints as given by Equations 9-1 through 9-3 are also imposed.

Example 9-3

Rework Example 9-2 using the Pedersen splitting correlation.

Solution

Step 1. Calculate coefficients A and B by the least-squares fit to the molar distribution of C_6 through C_{10} to give A = −14.404639 and B = −3.8125739.

Step 2. Solve for the mole fraction of C_{10} through C_{15} by applying Equation 9-6.

Component	Experimental z_n	Calculated z_n
C_7	0.000361	0.00361
C_8	0.00285	0.00285
C_9	0.00222	0.00222
C_{10}	0.00158	0.00166
C_{11}	0.00121	0.00128
C_{12}	0.00097	0.00098
C_{13}	0.00083	0.00076
C_{14}	0.00069	0.00058
C_{15}	0.00050	0.00045
C_{16+}	0.00094	0.00101*

*From Equation 9-1.

Ahmed's Method

Ahmed et al. (1985) [2] devised a simplified method for splitting the C_{7+} fraction into pseudo-components. The method originated from studying the molar behavior of 34 condensate and crude oil systems through detailed laboratory compositional analysis of the heavy fractions. The only required data for the proposed method are the molecular weight and the total mole fraction of the heptanes-plus fraction.

The splitting scheme is based on calculating the mole fraction z_n at a progressively higher number of carbon atoms. The extraction process continues until the sum of the mole fraction of the pseudo-components equals the total mole fraction of the heptanes-plus (z_{7+}).

$$z_n = z_{n+} \left[\frac{M_{(n+1)+} - M_{n+}}{M_{(n+1)+} - M_n} \right] \tag{9-7}$$

where z_n = mole fraction of the pseudo-component with a number of carbon atoms of n (z_7, z_8, z_9, etc.)

M_n = molecular weight of the hydrocarbon group with n carbon atoms

M_{n+} = molecular weight of the n+ fraction as calculated by the following expression:

$$M_{(n+1)+} = M_{7+} + S(n - 6) \qquad (9-8)$$

where n is the number of carbon atoms and S is the coefficient of Equation 9-8 with these values:

Number of Carbon Atoms	Condensate Systems	Crude Oil Systems
$n \leq 8$	15.5	16.5
$n > 8$	17.0	20.1

The stepwise calculation sequences of the proposed correlation are summarized in the following steps:

Step 1. According to the type of hydrocarbon system under investigation (condensate or crude oil), select appropriate values for the coefficients.

Step 2. Knowing the molecular weight of the C_{7+} fraction (M_{7+}), calculate the molecular weight of the octanes-plus fraction (M_{8+}) by applying Equation 9-8.

Step 3. Calculate the mole fraction of the heptane fraction (z_7) using Equation 9-7.

Step 4. Apply steps 2 and 3 repeatedly for each component in the system (C_8, C_9, etc.) until the sum of the calculated mole fractions is equal to the mole fraction of C_{7+} of the system.

The splitting scheme is best explained through the following example.

Example 9-4

Rework Example 9-3 using Ahmed's splitting method.

Solution

Step 1. Calculate the molecular weight of C_{8+} by applying Equation 9-8:

$$M_{8+} = 141.25 + 15.5(7 - 6) = 156.75$$

Step 2. Solve for the mole fraction of heptane (z_7) by applying Equation 9-7:

$$z_7 = z_{7+} \left[\frac{M_{8+} - M_{7+}}{M_{8+} - M_7} \right] = 0.0154 \left[\frac{156.75 - 141.25}{156.75 - 96} \right] = 0.00393$$

Step 3. Calculate the molecular weight of C_{9+} from Equation 9-8:

$$M_{9+} = 141.25 + 15.5(8 - 6) = 172.25$$

Step 4. Determine the mole fraction of C_8 from Equation 9-7:

$$z_8 = z_{8+}[(M_{9+} - M_{8+})/(M_{9+} - M_8)]$$
$$z_8 = (0.0154 - 0.00393)[(172.5 - 156.75)/(172.5 - 107)]$$
$$= 0.00276$$

Step 5. This extracting method is repeated as outlined in the preceding steps to give

Component	n	M_{n+}, Equation 9-8	M_n	z_n, Equation 9-7
C_7	7	141.25	96	0.000393
C_8	8	156.25	107	0.00276
C_9	9	175.25	121	0.00200
C_{10}	10	192.25	134	0.00144
C_{11}	11	209.25	147	0.00106
C_{12}	12	226.25	161	0.0008
C_{13}	13	243.25	175	0.00061
C_{14}	14	260.25	190	0.00048
C_{15}	15	277.25	206	0.00038
C_{16+}	16+	294.25	222	0.00159*

*Calculated from Equation 9-1.

Step 6. The boiling point, critical properties, and acentric factor of C_{16+} are then determined using the appropriate methods:

$$M = 222$$
$$\gamma = 0.856$$
$$T_b = 1174.6°R$$
$$P_c = 175.9 \, psia$$
$$T_c = 1449.3°R$$
$$\omega = 0.742$$

9.2 LUMPING SCHEMES

The large number of components necessary to describe the hydrocarbon mixture for accurate phase behavior modeling frequently burdens EOS calculations. Often, the problem is either lumping together the many experimentally determined fractions, or modeling the hydrocarbon system when the only experimental data available for the C_{7+} fraction are the molecular weight and specific gravity.

Generally, with a sufficiently large number of pseudo-components used in characterizing the heavy fraction of a hydrocarbon mixture, a satisfactory prediction of the PVT behavior by the equation of state can be obtained. However, in compositional models, the cost and computing time can increase significantly with the increased number of components in the system. Therefore, strict limitations are placed on the maximum number of components that can be used in compositional models and the original components have to be lumped into a smaller number of pseudo-components.

The term *lumping* or *pseudoization* then denotes the reduction in the number of components used in EOS calculations for reservoir fluids. This reduction is accomplished by employing the concept of the pseudo-component. The pseudo-component denotes a group of pure components lumped together and represented by a single component.

Several problems are associated with "regrouping" the original components into a smaller number without losing the predicting power of the equation of state. These problems include

- How to select the groups of pure components to be represented by one pseudo-component each.
- What mixing rules should be used for determining the EOS constants (p_c, T_c, and ω) for the new lumped pseudo-components.

A number of unique techniques have been published that can be used to address the lumping problems; notably the methods proposed by

- Lee et al. 1979 [30].
- Whitson (1980) [63].
- Mehra et al. (1983) [35].
- Montel and Gouel (1984) [36].
- Schlijper (1984) [50].
- Behrens and Sandler (1986) [5].
- Gouzalez et al. (1986) [16].

Three of these techniques are presented in the following discussion.

Whitson's Lumping Scheme

Whitson (1980) proposed a regrouping scheme whereby the compositional distribution of the C_{7+} fraction is reduced to only a few multiple-carbon-number (MCN) groups. Whitson suggested that the number of MCN groups necessary to describe the plus fraction is given by the following empirical rule:

$$N_g = \text{Int}[1 + 3.3 \log(N - n)] \tag{9-9}$$

where N_g = number of MCN groups

    Int = integer

    N = number of carbon atoms of the last component in the hydrocarbon system

    n = number of carbon atoms of the first component in the plus fraction, i.e., n = 7 for C_{7+}

The integer function requires that the real expression evaluated inside the brackets be rounded to the nearest integer. Whitson pointed out that, for black-oil systems, one could reduce the calculated value of N_g.

The molecular weights separating each MCN group are calculated from the following expression:

$$M_I = M_{C7}\left(\frac{M_{N+}}{M_{C7}}\right)^{1/N_g} \tag{9-10}$$

where $(M)_{N+}$ = molecular weight of the last reported component in the extended analysis of the hydrocarbon system:

$$M_{C7} = \text{molecular weight of } C_7$$
$$I = 1, 2, \ldots, N_g$$

Components with molecular weight falling within the boundaries of M_{I-1} to M_I are included in the Ith MCN group. Example 9-5 illustrates the use of Equations 9-9 and 9-10.

Example 9-5

Given the following compositional analysis of the C_{7+} fraction in a condensate system, determine the appropriate number of pseudo-components forming in the C_{7+}:

Component	z_i
C_7	0.00347
C_8	0.00268
C_9	0.00207
C_{10}	0.001596
C_{11}	0.00123
C_{12}	0.00095
C_{13}	0.00073
C_{14}	0.000566
C_{15}	0.000437
C_{16+}	0.001671

$M_{16+} = 259.$

Solution

Step 1. Determine the molecular weight of each component in the system:

Component	z_i	M_i
C_7	0.00347	96
C_8	0.00268	107
C_9	0.00207	121
C_{10}	0.001596	134
C_{11}	0.00123	147
C_{12}	0.00095	161
C_{13}	0.00073	175
C_{14}	0.000566	190
C_{15}	0.000437	206
C_{16+}	0.001671	259

Step 2. Calculate the number of pseudo-components from Equation 9-9:

$$N_g = \text{Int}[1 + 3.3\log(16 - 7)]$$
$$N_g = \text{Int}[4.15]$$
$$N_g = 4$$

Step 3. Determine the molecular weights separating the hydrocarbon groups by applying Equation 9-10:

$$M_I = 96\left[\frac{259}{96}\right]^{1/4}$$
$$M_I = 96[2.698]^{1/4}$$

I	$(M)_I$
1	123
2	158
3	202
4	259

- **First pseudo-component**. The first pseudo-component includes all components with molecular weight in the range of 96 to 123. This group then includes C_7, C_8, and C_9.
- **Second pseudo-component**. The second pseudo-component contains all components with a molecular weight higher than 123 to a molecular weight of 158. This group includes C_{10} and C_{11}.

- **Third pseudo-component**. The third pseudo-component includes components with a molecular weight higher than 158 to a molecular weight of 202. Therefore, this group includes C_{12}, C_{13}, and C_{14}.
- **Fourth pseudo-component**. This pseudo-component includes all the remaining components, i.e., C_{15} and C_{16+}.

Group I	Component	z_i	z_I
1	C_7	0.00347	0.00822
	C_8	0.00268	
	C_9	0.00207	
2	C_{10}	0.001596	0.002826
	C_{11}	0.00123	
3	C_{12}	0.00095	0.002246
	C_{13}	0.00073	
	C_{14}	0.000566	
4	C_{15}	0.000437	0.002108
	C_{16+}	0.001671	

It is convenient at this stage to present the mixing rules that can be employed to characterize the pseudo-component in terms of its pseudo-physical and pseudo-critical properties. Because there are numerous ways to mix the properties of the individual components, all giving different properties for the pseudo-components, the choice of a correct mixing rule is as important as the lumping scheme. Some of these mixing rules are given next.

Hong's Mixing Rules

Hong (1982) [22] concluded that the weight fraction average w_i is the best mixing parameter in characterizing the C_{7+} fractions by the following mixing rules:

- Pseudo-critical pressure $p_{cL} = \sum^{L} w_i p_{ci}$.
- Pseudo-critical temperature $T_{cL} = \sum^{L} w_i T_{ci}$.
- Pseudo-critical volume $V_{cL} = \sum^{L} w_i V_{ci}$.
- Pseudo-acentric factor $\omega_L = \sum^{L} [\phi_i \omega_i]$.
- Pseudo-molecular weight $M_L = \sum^{L} w_i M_i$.
- Binary interaction coefficient $K_{kL} = 1 - \sum_i^{L} \sum_j^{L} w_i w_j (1 - k_{ij})$.

with

$$w_i = \frac{z_i M_i}{\sum^{L} z_i M_i}$$

where w_i = average weight fraction

K_{kL} = binary interaction coefficient between the kth component and the lumped fraction

The subscript L in this relationship denotes the lumped fraction.

Lee's Mixing Rules

Lee et al. 1979 [30], in their proposed regrouping model, employed Kay's mixing rules as the characterizing approach for determining the properties of the lumped fractions. Defining the normalized mole fraction of the component i in the lumped fraction as

$$\phi_i = z_i \bigg/ \sum^{L} z_i$$

the following rules are proposed:

$$M_L = \sum^{L} \phi_i M_i \tag{9-11}$$

$$\gamma_L = M_L \bigg/ \sum^{L} [\phi_i M_i / \gamma_i] \tag{9-12}$$

$$V_{cL} = \sum^{L} [\phi_i M_i V_{ci} / M_L] \tag{9-13}$$

$$P_{cL} = \sum^{L} [\phi_i P_{ci}] \tag{9-14}$$

$$T_{cL} = \sum^{L} [\phi_i T_{ci}] \tag{9-15}$$

$$\omega_L = \sum^{L} [\phi_i \omega_i] \tag{9-16}$$

Example 9-6

Using Lee's mixing rules, determine the physical and critical properties of the four pseudo-components in Example 9-5.

Solution

Step 1. Assign the appropriate physical and critical properties to each component:

Group	Comp.	z_i	I	M_i	γ_i	V_{ci}	p_{ci}	T_{ci}	ω_i
1	C_7	0.00347		96	0.272	0.06289	453	985	0.280
	C_8	0.00268	0.00822	107	0.748	0.06264	419	1036	0.312
	C_9	0.00207		121	0.768	0.06258	383	1058	0.348
2	C_{10}	0.001596	0.002826	134	0.782	0.06273	351	1128	0.385
	C_{11}	0.00123		147	0.793	0.06291	325	1166	0.419
3	C_{12}	0.00095		161	0.804	0.06306	302	1203	0.454
	C_{13}	0.00073	0.002246	175	0.815	0.06311	286	1236	0.484
	C_{14}	0.000566		190	0.826	0.06316	270	1270	0.516
4	C_{15}	0.000437	0.002108	206	0.826	0.06325	255	1304	0.550
	C_{16+}	0.001671		259	0.908	0.0638^*	215^*	1467	0.68^*

*Calculated.

Step 2. Calculate the physical and critical properties of each group by applying Equations 9-11 through 9-16 to give

Group	Z_I	M_L	γ_L	V_{cL}	p_{cL}	T_{cL}	ω_L
1	0.00822	105.9	0.746	0.0627	424	1020	0.3076
2	0.002826	139.7	0.787	0.0628	339.7	1144.5	0.4000
3	0.002246	172.9	0.814	0.0631	288	1230.6	0.4794
4	0.002108	248	0.892	0.0637	223.3	1433	0.6531

Problems

1. A hydrocarbon system has the following composition:

Component	z_i
C_1	0.30
C_2	0.10
C_3	0.05
C_4	0.03
nC_4	0.03
C_5	0.02
nC_5	0.02
C_6	0.05
C_{7+}	0.40

Given the following additional data:

$$\text{System pressure} = 2100 \text{ psia}$$
$$\text{System temperature} = 150°F$$
$$\text{Specific gravity of } C_{7+} = 0.80$$
$$\text{Molecular weight of } C_{7+} = 140$$

calculate the equilibrium ratios of the above system.

2. A well is producing oil and gas with the following compositions at a gas-oil ratio of 500 scf/STB:

Component	x_i	y_i
C_1	0.35	0.60
C_2	0.08	0.10
C_3	0.07	0.10
$n-C_4$	0.06	0.07
$n-C_5$	0.05	0.05
C_6	0.05	0.05
C_{7+}	0.34	0.05

Given the following additional data:

$$\text{Current reservoir pressure} = 3000 \text{ psia}$$
$$\text{Bubble-point pressure} = 2800 \text{ psia}$$
$$\text{Reservoir temperature} = 120°F$$

$$M \text{ of } C_{7+} = 125$$
$$\text{Specific gravity of } C_{7+} = 0.823$$

calculate the composition of the reservoir fluid.

3. A saturated hydrocarbon mixture with the following composition exists in a reservoir at 234°F:

Component	z_i
C_1	0.3805
C_2	0.0933
C_3	0.0885
C_4	0.0600
C_5	0.0378
C_6	0.0356
C_{7+}	0.3043

Calculate

a. The bubble-point pressure of the mixture.
b. The compositions of the two phases if the mixture is flashed at 500 psia and 150°F.
c. The density of the liquid phase.
d. The compositions of the two phases if the liquid from the first separator is further flashed at 14.7 psia and 60°F.
e. The oil formation volume factor at the bubble-point pressure.
f. The original gas solubility.
g. The oil viscosity at the bubble-point pressure.

4. A crude oil exists in a reservoir at its bubble-point pressure of 2520 psig and a temperature of 180 °F. The oil has the following composition:

Component	x_i
CO_2	0.0044
N_2	0.0045
C_1	0.3505
C_2	0.0464
C_3	0.0246
$i-C_4$	0.0683
$n-C_4$	0.0083
$i-C_5$	0.0080
$n-C_5$	0.0080
C_6	0.0546
C_{7+}	0.4824

The molecular weight and specific gravity of C_{7+} are 225 and 0.8364. The reservoir contains initially 12 MMbbl of oil. The surface facilities consist of two separation stages connecting in series. The first separation stage operates at 500 psig and 100°F. The second stage operates under standard conditions.

a. Characterize C_{7+} in terms of its critical properties, boiling point, and acentric factor.
b. Calculate the initial oil in place in STB.
c. Calculate the standard cubic feet of gas initially in solution.
d. Calculate the composition of the free gas and the composition of the remaining oil at 2495 psig, assuming the overall composition of the system will remain constant.

5. A pure n-butane exists in the two-phase region at 120°F. Calculate the density of the coexisting phase using the following equations of state:
a. Van der Waals
b. Redlich-Kwong
c. Soave-Redlich-Kwong
d. Peng-Robinson

6. A crude oil system with the following composition exists at its bubble-point pressure of 3,250 psia and 155°F:

Component	x_i
C_1	0.42
C_2	0.08
C_3	0.06
C_4	0.02
C_5	0.01
C_6	0.04
C_{7+}	0.37

If the molecular weight and specific gravity of the heptanes-plus fraction are 225 and 0.823, respectively, calculate the density of the crude oil using

a. Van der Waals EOS
b. Redlich-Kwong EOS
c. SRR EOS
d. PR EOS

7. Calculate the vapor pressure of propane at 100 °F using
a. Van der Waals EOS
b. SRK EOS
c. PR EOS

Compare the results with that obtained from the Cox chart.

8. A natural gas exists at 2000 psi and 150°F. The gas has the following composition:

Component	y_i
C_1	0.80
C_2	0.10
C_3	0.07
$i - C_4$	0.02
$n - C_4$	0.01

Calculate the density of the gas using the following equations of state:

a. Van der Waals
b. RK
c. SRK
d. PR

9. The heptanes-plus fraction in a condensate gas system is character-ized by a molecular weight and specific gravity of 190 and 0.8, respectively. The mole fraction of the C_{7+} is 0.12. Extend the molar distribution of the plus fraction to C_{20+} using
a. Katz's method.
b. Ahmed's method.

Determine the critical properties of C_{20+}.

10. A naturally occurring crude oil system has a heptanes-plus fraction with the following properties:

$$M_{7+} = 213$$
$$\gamma_{7+} = 0.8405$$
$$x_{7+} = 0.3497$$

Extend the molar distribution of the plus fraction to C_{25+} and deter-mine the critical properties and acentric factor of the last component.

11. A crude oil system has the following composition:

Component	x_i
C_1	0.3100
C_2	0.1042
C_3	0.1187
C_4	0.0732
C_5	0.0441
C_6	0.0255
C_{7+}	0.3243

The molecular weight and specific gravity of C_{7+} are 215 and 0.84, respectively.

a. Extend the molar distribution of C_{7+} to C_{20+}.

b. Calculate the appropriate number of pseudo-components necessary to adequately represent the composition from C_7 to C_{20+} and characterize the resulting pseudo-components in terms of

- Molecular weight
- Specific gravity
- Critical properties
- Acentric factor

References

[1] T. Ahmed, A Practical Equation of State, SPERE 291 (1991) 136–137.

[2] T. Ahmed, G. Cady, A. Story, A Generalized Correlation for Characterizing the Hydrocarbon Heavy Fractions, in: SPE Paper 14266 presented at the SPE 60th Annual Technical Conference, Las Vegas, NV, 1985.

[3] G.H. Alani, H.T. Kennedy, Volume of Liquid Hydrocarbons at High Temperatures and Pressures, Trans. AIME 219 (1960) 288–292.

[4] J. Amyx, D. Bass, R. Whitney, Petroleum Reservoir Engineering, McGraw-Hill Book Company, New York, 1960.

[5] R. Behrens, S. Sandler, The Use of Semi-Continuous Description to Model the C_{7+} Fraction in Equation of State Calculation, in: SPE/DOE Paper 14925 presented at the 5th Annual Symposium on EOR, Tulsa, OK, 1986.

[6] F.H. Brinkman, J.N. Sicking, Equilibrium Ratios for Reservoir Studies, Trans. AIME 219 (1960) 313–319.

[7] J.M. Campbell, Gas Conditioning and Processing, vol. 1, Campbell Petroleum Series, Norman, OK, 1976.

[8] P. Chueh, J. Prausnitz, Vapor-Liquid Equilibria at High Pressures: Calculation of Critical Temperatures, Volumes, and Pressures of Nonpolar Mixtures, AIChE Journal 13 (6) (1967) 1107–1112.

[9] N. Clark, A Review of Reservoir Engineering, World Oil (1951).

[10] N. Clark, Elements of Petroleum Reservoirs, Society of Petroleum Engineers, Dallas, 1960.

[11] H. Dykstra, T.D. Mueller, Calculation of Phase Composition and Properties for Lean- or Enriched-Gas Drive, SPEJ (1965) 239–246.

[12] W.C. Edmister, Applied Hydrocarbon Thermodynamics, Part 4, Compressibility Factors and Equations of State, Petroleum Refiner 37 (1958) 173–179.

[13] W. Edmister, B. Lee, Applied Hydrocarbon Thermodynamics, vol. 1, second ed., Gulf Publishing Company, Houston, 1986.

[14] J. Elliot, T. Daubert, Revised Procedure for Phase Equilibrium Calculations with Soave Equation of State, Ind. Eng. Chem. Process Des. Dev. 23 (1985) 743–748.

[15] R. Gibbons, A. Laughton, An Equation of State for Polar and Non-Polar Substances and Mixtures, J. Chem. Soc. 80 (1984) 1019–1038.

[16] E. Gonzalez, P. Colonomos, I. Rusinek, A New Approach for Characterizing Oil Fractions and for Selecting Pseudo-Components of Hydrocarbons, Canadian JPT (1986) 78–84.

[17] M.S. Graboski, T.E. Daubert, A Modified Soave Equation of State for Phase Equilibrium Calculations 1. Hydrocarbon System, Ind. Eng. Chem. Process Des. Dev. 17 (1978) 443–448.

[18] J.T. Hadden, Convergence Pressure in Hydrocarbon Vapor-Liquid Equilibria, Chem. Eng. Progr. Symposium Ser. 49 (7) (1953) 53.

[19] O. Hariu, R. Sage, Crude Split Figured by Computer, Hydrocarbon Process (1969) 143–148.

[20] G. Heyen, A Cubic Equation of State with Extended Range of Application, in: Paper presented at 2nd World Congress Chemical Engineering, Montreal, 1983.

[21] A.E. Hoffmann, J.S. Crump, R.C. Hocott, Equilibrium Constants for a Gas-Condensate System, Trans. AIME 198 (1953) 1–10.

[22] K.C. Hong, Lumped-Component Characterization of Crude Oils for Compositional Simulation, in: SPE/DOE Paper 10691 presented at the 3rd Joint Symposium on EOR, Tulsa, OK, 1982.

[23] B.S. Jhaveri, G.K. Youngren, Three-Parameter Modification of the Peng-Robinson Equation of State to Improve Volumetric Predictions, in: SPE Paper 13118 presented at the 1984 SPE Annual Technical Conference, Houston, September 16–19.

[24] D.L. Katz, K.H. Hachmuth, Vaporization Equilibrium Constants in a Crude Oil–Natural Gas System, Ind. Eng. Chem. 29 (1937) 1072.

[25] D. Katz, et al., Handbook of Natural Gas Engineering, McGraw-Hill Book Company, New York, 1959.

[26] D. Katz, et al., Overview of Phase Behavior of Oil and Gas Production, JPT (1983) 1205–1214.

[27] D.M. Kehn, Rapid Analysis of Condensate Systems by Chromatography, JPT (1964) 435–440.

[28] W.L.J. Kubic, A Modification of the Martin Equation of State for Calculating Vapor-Liquid Equilibria, Fluid Phase Equilibria 9 (1982) 79–97.

[29] B. Lee, G. Kesler, A Generalized Thermodynamic Correlation Based on Three-Parameter Corresponding States, AIChE Journal 21 (3) (1975) 510–527.

[30] S. Lee, et al., Experimental and Theoretical Studies on the Fluid Properties Required for Simulation of Thermal Processes, in: SPE Paper 8393 presented at the SPE 54th Annual Technical Conference, Las Vegas, NV, 1979.

[31] D. Lim, et al., Calculation of Liquid Dropout for Systems Containing Water, in: SPE Paper 13094 presented at the SPE 59th Annual Technical Conference, Houston, 1984.

[32] J. Lohrenz, B.G. Bray, C.R. Clark, Calculating Viscosities of Reservoir Fluids from their Compositions, JPT (1964) 1171, Trans. AIME 231.

[33] R.N. Maddox, J.H. Erbar, Gas Conditioning and Processing, Vol. 3—Advanced Techniques and Applications, Campbell Petroleum Series, Norman, OK, 1982.

[34] R.N. Maddox, J.H. Erbar, Improve P-V-T Predictions, Hydrocarbon Process. (1984) 119–121.

[35] R. Mehra, et al., A Statistical Approach for Combining Reservoir Fluids into Pseudo Components for Compositional Model Studies, in: SPE Paper 11201 presented at the SPE 57th Annual Meeting, New Orleans, 1983.

[36] F. Montel, P. Gouel, A New Lumping Scheme of Analytical Data for Composition Studies, in: SPE Paper 13119 presented at the SPE 59th Annual Technical Conference, Houston, 1984.

[37] V. Nikos, et al., Phase Behavior of Systems Comprising North Sea Reservoir Fluids and Injection Gases, JPT (1986) 1221–1233.

[38] N. Patel, A. Teja, A New Equation of State for Fluids and Fluid Mixtures, Chem. Eng. Sci. 37 (3) (1982) 463–473.

[39] K. Pedersen, P. Thomassen, A. Fredenslund, Phase Equilibria and Separation Processes, Report SEP 8207, Institute for Kemiteknik, Denmark Tekniske Hojskole, Denmark, 1982.

[40] A. Peneloux, E. Rauzy, R. Freze, A Consistent Correlation for Redlich–Kwong–Soave Volumes, Fluid Phase Equilibria 8 (1982) 7–23.

[41] D. Peng, D. Robinson, A New Two Constant Equation of State, Ind. Eng. Chem. Fund. 15 (1) (1976a) 59–64.

[42] D. Peng, D. Robinson, Two and Three Phase Equilibrium Calculations for Systems Containing Water, Canadian J. Chem. Eng. 54 (1976b) 595–598.

[43] D. Peng, D. Robinson, Two and Three Phase Equilibrium Calculations for Coal Gasification and Related Processes, in: ACS Symposium Series, No. 133, Thermodynamics of Aqueous Systems with Industrial Applications, American Chemical Society, Washington, DC, 1980.

[44] C.S. Peterson, A Systematic and Consistent Approach to Determine Binary Interaction Coefficients for the Peng-Robinson Equations of State, SPERE (1989) 488–496.

[45] O. Redlich, J. Kwong, On the Thermodynamics of Solutions. An Equation of State. Fugacities of Gaseous Solutions, Chem. Rev. 44 (1949) 233–247.

[46] R. Reid, J.M. Prausnitz, T. Sherwood, The Properties of Gases and Liquids, third ed., McGraw-Hill Book Company, New York, 1977.

[47] M.R. Riazi, T.E. Daubert, Characterization Parameters for Petroleum Fractions, Ind. Eng. Chem. Res. 26 (24) (1987) 755–759.

[48] D.B. Robinson, D.Y. Peng, The Characterization of the Heptanes and Heavier Fractions, Research Report 28, GPA, Tulsa, 1978.

[49] M.J. Rzasa, E.D. Glass, J.B. Opfell, Prediction of Critical Properties and Equilibrium Vaporization Constants for Complex Hydrocarbon Systems, Chem. Eng. Progr. Symposium Ser. 48 (2) (1952) 28.

[50] A.G. Schlijper, Simulation of Compositional Process: The Use of Pseudo-Components in Equation of State Calculations, in: SPE/DOE Paper 12633 presented at the SPE/DOE 4th Symposium on EOR, Tulsa, OK, 1984.

[51] G. Schmidt, H. Wenzel, A Modified Van der Waals Type Equation of State, Chem. Eng. Sci. 135 (1980) 1503–1512.

[52] W.J. Sim, T.E. Daubert, Prediction of Vapor-Liquid Equilibria of Undefined Mixtures, Ind. Eng. Chem. Process Des. Dev. 19 (3) (1980) 380–393.

[53] C. Slot-Petersen, A Systematic and Consistent Approach to Determine Binary Interaction Coefficients for the Peng–Robinson Equation of State, in: SPE Paper 16941 presented at the SPE 62 Annual Technical Conference, Dallas, 1987.

[54] G. Soave, Equilibrium Constants from a Modified Redlich–Kwong Equation of State, Chem. Eng. Sci. 27 (1972) 1197–1203.

[55] C. Spencer, T. Daubert, R. Danner, A Critical Review of Correlations for the Critical Properties of Defined Mixtures, AIChE Journal 19 (3) (1973) 522–527.

[56] M.B. Standing, Volumetric and Phase Behavior of Oil Field Hydrocarbon Systems, Society of Petroleum Engineers of AIME, Dallas, 1977.

[57] M.B. Standing, A Set of Equations for Computing Equilibrium Ratios of a Crude Oil/Natural Gas System at Pressures Below 1,000 psia, JPT (1979) 1193–1195.

[58] M.B. Standing, D.L. Katz, Density of Crude Oils Saturated with Natural Gas, Trans. AIME 146 (1942) 159–165.

[59] R. Stryjek, J.H. Vera, PRSV: An Improvement to the Peng-Robinson Equation of State for Pure Compounds and Mixtures, Canadian J. Chem. Eng. 64 (1986) 323–333.

[60] J. Valderrama, L. Cisternas, A Cubic Equation of State for Polar and Other Complex Mixtures, Fluid Phase Equilibria 29 (1986) 431–438.

[61] J.D. Van der Waals, On the Continuity of the Liquid and Gaseous State, Ph.D. Dissertation, Sigthoff, Leiden, (1873).

[62] J. Vidal, T. Daubert, Equations of State—Reworking the Old Forms, Chem. Eng. Sci. 33 (1978) 787–791.

[63] C. Whitson, Characterizing Hydrocarbon Plus Fractions, in: EUR Paper 183 presented at the European Offshore Petroleum Conference, London, 1980.

[64] C. Whitson, M. Brule, Phase Behavior, Society of Petroleum Engineers, Inc., Richardson, TX, 2000.

[65] C.H. Whitson, S.B. Torp, Evaluating Constant Volume Depletion Data, in: SPE Paper 10067 presented at the SPE 56th Annual Fall Technical Conference, San Antonio, TX, 1981.

[66] B.T. Willman, B.T. Teja, Continuous Thermodynamics of Phase Equilibria Using a Multivariable Distribution Function and an Equation of State, AIChE Journal 32 (12) (1986) 2067–2078.

[67] G. Wilson, A Modified Redlich–Kwong EOS, Application to General Physical Data Calculations, in: Paper 15C presented at the Annual AIChE National Meeting, Cleveland, OH, 1968.

[68] F.W. Winn, Simplified Nomographic Presentation, Hydrocarbon Vapor–Liquid Equilibria, Chem. Eng. Progr. Symposium Ser. 33 (6) (1954) 131–135.

Index

A
Ahmed's method, 118–119
Alani–Kennedy method
  apparent molecular weight, 56
  coefficients, 55–56, 55t
  crude oil density, 57
  heptanes-plus, 55–56
  hydrocarbon mixture density, 56
  molar volume calculation, 54, 56
  van der Waals equation, 54–55

B
Bubble-point pressure
  petroleum engineering, 101–102
  reservoir engineering, 32–33

C
Campbell's method, 25
Convergence pressure method
  graphical correlation, 20
  Hadden's method, 20–22
  hydrocarbon mixture, 18, 19
  Natural Gas Processors Suppliers
    Association (NGPSA), 18–19, 20
  Rzasa's method, 22
  Standing's method, 22
Cox chart, 2–3, 2f

D
Dalton's law, 5, 13
Dew-point pressure
  petroleum engineering
    computational procedure, 100–101
    fugacity coefficient, 99, 100
  reservoir engineering, 29–31

E
Edmister correlation, 81
Equation of state (EOS)
  lumping schemes, 120, 121
  Peng–Robinson equation of state
    acentric values, 86
    binary interaction coefficient, 94–96

compressibility factor, 85–86
corrected hydrocarbon phase
  volumes, 92
fugacity, 87–88
fugacity coefficient, 88
heptanes-plus fraction, 93, 94
liquid density prediction, 85
nonlinear regression model, 93, 94
phase and volumetric behavior, 93
pressure correction, 90
Riazi and Daubert correlation, 93
shift parameter, 92–93
temperature-dependent parameter,
  85–86
vapor pressure reproduction, 90–91
volume correction parameter, 91–92
petroleum engineering
  bubble-point pressure, 101–102
  dew-point pressure, 98–101
  equilibrium ratios, 97–98
  three-phase equilibrium calculations,
    102–107
  vapor pressure, 107–109
Redlich–Kwong equation of state, 66–71
Soave–Redlich–Kwong equation of state
  acentric factor, 72, 81
  binary interaction coefficient, 75, 76,
    81–82
  compressibility factor, 73, 76, 82
  correction factor, 83
  critical isotherm, 72
  dimensionless pure component
    parameters, 72
  Edmister correlation, 81
  equilibrium ratio, 80
  fugacity, 78–79
  fugacity coefficient, 79, 80–81
  Lee and Kessler correction, 81
  liquid and gas phase, 76
  molar volume, 72–73
  temperature-dependent term $(a\alpha)$, 71–72
  thermodynamic equilibrium, 79–80

Printed in the United States
By Bookmasters